OCEANUS
The Marine Environment

Study Guide

Ninth Edition

Tom Garrison
Ruth Lebow

BROOKS/COLE

THOMSON LEARNING

Australia • Canada • Mexico • Singapore • Spain
United Kingdom • United States

For permission to use material from this text,
contact us by Web: http://www.thomsonrights.com
Fax: 1-800-730-2215 Phone: 1-800-730-2214

Printer: Patterson Printing

ISBN 0-534-37563-4

For more information, contact
Wadsworth/Thomson Learning
10 Davis Drive
Belmont, CA 94002-3098
USA

For more information about our products, contact us:
Thomson Learning Academic Resource Center
1-800-423-0563
http://www.wadsworth.com

International Headquarters
Thomson Learning
International Division
290 Harbor Drive, 2nd Floor
Stamford, CT 06902-7477
USA

UK/Europe/Middle East/South Africa
Thomson Learning
Berkshire House
168-173 High Holborn
London WC1V 7AA
United Kingdom

Asia
Thomson Learning
60 Albert Complex, #15-01
Singapore 189969

Canada
Nelson Thomson Learning
1120 Birchmount Road
Toronto, Ontario M1K 5G4
Canada

Contents

Preface v

Introduction vii

Lesson 1 The Water Planet 1

Lesson 2 Cosmic Origins 6

Lesson 3 Historical Perspectives 11

Lesson 4 The Waters of the Earth 19

Lesson 5 Ocean's Edge 25

Lesson 6 The Intertidal Zone 30

Lesson 7 Continental Margins 35

Lesson 8 Beyond Land's End 41

Lesson 9 Plate Tectonics 47

Lesson 10 Islands 53

Lesson 11 Marine Meteorology 58

Lesson 12 Ocean Currents 65

Lesson 13 Wind Waves and Water Dynamics . 71

Lesson 14 The Ebb and Flow 77

Lesson 15 Plankton: Floaters and Drifters 83

Lesson 16 Nekton: Swimmers 89

Lesson 17 Reptiles and Birds 94

Lesson 18 Mammals: Seals and Otters 99

Lesson 19 Mammals: Whales 104

Lesson 20 Living Together 111

Lesson 21 Light in the Sea 116

Lesson 22 Sound in the Sea 121

Lesson 23 Life Under Pressure 127

Lesson 24 The Polar Seas 133

Lesson 25 The Tropic Seas 140

Lesson 26 Mineral Resources 146

Lesson 27 Biological Resources 152

Lesson 28 Marine Pollution 159

Lesson 29 Hawaii: A Case Study 167

Lesson 30 Epilogue 173

Answer Key for the Self-Tests 175

Preface

And I have loved thee, Ocean! and my joy
Of youthful sports was on thy breast to be
Borne like thy bubbles, onward: from a boy
I wanton'd with thy breakers – they to me
Were a delight; and if the freshening sea
Made them a terror – 'twas a pleasing far,
For I was as it were a child of thee,
And trusted to thy billows for and near,
And laid my hand upon thy mane – as I do here.

George Gordon,
Lord Byron

Few courses of study are more intriguing to life-long learners or undergraduates attempting to fulfill their general education science requirement than introductory oceanography. The visual and intellectual richness of the marine sciences makes a deep impression on students and contributes significantly to their awareness and appreciation of the natural environment. Few of us can remain impassive while viewing film of a tidal wave striking the Hawaiian Islands, reading an account of the U.S. Navy's encounter with the largest wave ever measured at sea, considering the origin of the molecules that make up our bodies, reading about the amazing adventures of polar explorers such as Ernest Shackleton, or hearing the lovely sound of dolphins and whales communicating.

Whether you live in a coastal city or an inland town, the ocean affects you. It affects you because of its influence on the earth's weather, its stunning physical size and diversity of life forms, its contributions to the physical and historical development of humans, its impact on geopolitical and economic matters, and its importance in literature and the graphic arts.

The goals of this course focus on the uniqueness of the earth and on areas of intense scientific and public concern: plate tectonics and earthquake prediction, the impact of oceanic pollutants, climatic fluctuations, and the potential exploitation of marine resources. If, as microbiologist Barry Commoner predicts, we have "perhaps a generation in which to save the environment from the final effects of the violence we have done it," what better use could we make of our time than learning about the water environment that covers 70.8% of the earth on which we live?

Introduction

This is your study guide. Treasure it; struggle with it; use it as your personal road map to the world of *Oceanus*. Following the instructions in this guide will enable you to devote your time to absorbing course content rather than puzzling over what it is you are supposed to be learning.

Whether you are a newcomer to television courses, or an old hand, it is important that you become familiar with the components of *Oceanus*, how they function with each other, and how they relate to you. Each of the elements in a telecourse contributes to the whole, effectively using the style of communication peculiar to that particular medium.

The Television Lesson

The video component of *Oceanus* calls your attention to key concepts and abstract ideas through a variety of formats, from documentary to demonstration. The camera has the ability to travel and observe many activities not normally available to the classroom student. It brings noted authorities into your home on a moment's notice. You become an eyewitness to natural phenomena rarely seen in a lifetime.

Using television for learning is not like watching a comedy series or sporting event. At first you will have to concentrate on *active* watching. It is very easy to slip into the passive, half-viewing stance used for entertainment television. In most instances, you will have a chance to review the lesson in an alternate time period, or watch video cassettes of the lesson at the learning center on campus.

If you have an audio recorder available, tape the audio portion of the program as you view it. After you have watched the program and can visualize it, the audio portion is an excellent source for review. If you have any questions about content or wish additional information, contact the faculty advisor at the campus where you are enrolled. He or she is eager to help.

The Text

Oceanus is keyed to two textbooks, either of which may be used with the television lessons: *Oceanography, An Invitation to Marine Science*, fourth edition, by Tom Garrison, published by Wadsworth in 2002; or *Oceanography: An Introduction*, Fifth Edition, by Dale Ingmanson and William Wallace, published by Wadsworth in 1995. A text is an essential part of the course, providing information most appropriate to the print medium. It establishes a foundation of knowledge and elaborates on concepts introduced in the television segments through charts, studies, research, and additional photographs. The assignment section of the study guide coordinates reading and review assignments with the television segment.

You should leaf through each assignment before you begin to read. Some text assignments are fairly short, some rather long. In this way you can gauge the length and detail of the reading, and pace yourself.

Your campus instructor will tell you which of the texts to use. Assignments will be given for both books (you need not read both).

The Study Guide

As the title indicates, the study guide helps you to synthesize and integrate the materials presented in the text and television segments. Each lesson of the study guide contains the following:

☐ The *Overview* is a summary of the highlights of that particular lesson – major facts, concepts, and opinions placed in the total perspective of the course.

☐ The *Learning Objectives* are the goals you are expected to accomplish as a result of completing the required activities for that particular lesson.

☐ The *Key Terms and Phrases* segment provides a vocabulary list of important words and expressions that will be encountered in the lesson. This segment complements but does not substitute for the Glossary in the textbook.

☐ *Before Viewing* informs you of the required reading for that particular lesson and may direct you to perform some type of required activity in preparation for viewing the television program.

☐ The *After Viewing* section includes a variety of opportunities to apply and practice what you have learned from the text and video lessons. These activities are required. Sometimes answers are provided; in other instances you will be given the chance to plot your solution and test your knowledge without help. Any questions, discoveries, or difficulties may be shared with your campus instructor.

☐ The *Optional Activities* section suggests activities to expand your understanding of the subject matter.

☐ The *Self-Test* is a series of multiple-choice questions designed to serve two purposes: to test your understanding of the objectives of each lesson and to prepare you to take the course examinations.

☐ The *Supplemental Readings* is a list of suggested books and articles for additional information on the various concepts and issues discussed in the lesson.

Occasionally, other special sections and notes that are specific to a particular lesson will be added for clarification or to identify supplementary resources.

As we indicated earlier, each component in a telecourse contributes to the whole. The course is not any one of the elements by itself. It is a blending of all three: the television lesson, the text, and the study guide.

How to Use This Study Guide

These few suggestions may assist you in using this study guide and in understanding the rationale for its design.

First, read the Overview, study the Learning Objectives and Key Terms and Phrases, and do all the Before Viewing assignments for each lesson before you watch the corresponding television program. Each 30-minute television episode will go by too quickly for you to take comprehensive notes. Moreover, if you try to take notes on everything, you will soon discover that you have missed a great deal of the visual impact of the program because you are looking at a pad of paper. By completing all the pre-viewing requirements before watching the program, you are preparing yourself to select and retain those major concepts and facts not fully covered in the text. You will already have acquired the specialized vocabulary needed to understand the content of the television program, thereby eliminating one more obstacle in the learning process, and you will gain the ability to distinguish between that which is truly important and that which is interesting, but incidental. Test items may cover material presented in the video program only, so be sure to pay attention and review the programs, if possible, on audio or videocassette.

Second, it is a good idea to reread the assigned passages in the text as soon as possible after viewing the television lesson to reinforce what you have just seen. The activities in the After Viewing section are designed to further your understanding of the objectives for each lesson, so they should be taken seriously. Incidentally, inasmuch as the After Viewing portion of the study sequence is required, you may be tested on the content.

Third, take the Self-Test honestly. Make a genuine attempt to answer each question without glancing first at the Answer Key. The Self-Test is your opportunity to practice what you have just learned, without penalty, and with almost instantaneous feedback. Make the most of this opportunity. Remember, the Self-Tests are a dry run for the examinations to come. Whenever you miss a question on the Self-Test, take the time to find out why. Use your errors to find out what fact or concept you missed or misunderstood along the way. You can learn as much from a wrong answer as from a right one and can avoid making the same mistake again.

Fourth, if you should experience difficulty, seek help from the campus instructor. All of us who were involved in designing and producing this course hope not only that you will learn a great deal about the ocean environment, but also that the whole process of learning will be an exciting and enjoyable one.

Lesson One   The Water Planet

Overview

The Greeks were experienced voyagers and knew their Mediterranean well. Occasionally they would venture past the protecting arms of the Straits of Gibraltar and encounter the "ocean-river," a great breathing and moving entity they named Oceanus. This vast, boundless river was believed to flow endlessly around the earth. Homer, with remarkable prescience, named Oceanus as the origin of all things, including the gods. In Hesiod's *Theogony*, Oceanus is a Titan, son of Uranus and Gaea (Heaven and Earth). Oceanus and his wife Tethys, according to mythology, bore 3,000 sons (the rivers) and 3,000 daughters (the water nymphs, Oceanides).

A Greek view of the ocean-river is an altogether appropriate way to begin our work. During the childhood of the world thoughtful Greeks were as awed by the ocean and its creatures as we are today. Through their mythology comes a sense of the beauty, power, majesty, and potential of the seas. The astronauts have spoken lovingly of the sight of this ocean and the world it dominates. Surely it is as lovely a sight from space as it was, and is, from the temples of Poseidon.

Earth is unique. Seventy and eight-tenths percent of the planet is covered by water. The oceans affect and moderate temperature, dramatically influence weather, provide at least two percent of the world's direct human intake of protein, support about one-third of the world's petroleum demand, border most of the world's largest cities, and supply primary shipping and communication routes and major recreational resources. We will someday look to the seas for additional food, raw materials, energy and energy resources, living space, and special products.

The ocean also lives within us. Whether terrestrial, aquatic, or marine, all life evolved in the ocean, developing and flourishing there for a few billion years before venturing onto the unwelcoming land. All life on earth depends on an internal seawater mix to provide a suitable environment for life processes. The life forms of the land carry an ocean within themselves. Their blood, their eggs, even the fluids that bathe their cells are oceanic. Humans are at least 71 percent water – water that contains nearly the same concentration and proportion of major elements as the oceans themselves. The first nine months of human life are spent in a water world – a warm, supportive ocean that cradles from shock and provides a stable and weightless environment for the complex processes of growth and development. After birth, our entire personal view of the universe is seen through an ocean: The fluid behind the cornea in our eyes is a very close analog to seawater. Without water, life as we know it would be impossible. Cells require water as surely as they require oxygen and food.

There is also a diversity and potential for intelligence among oceanic life that is likewise unmatched by most terrestrial counterparts. This seems plausible when one understands that modern scientific theory supports the premise that life evolved in the ocean. Though the seas were much different 3.5 to 4 billion years ago, all of the elements necessary for organic molecules were present in those ancient waters. Scientific investigation has demonstrated that amino acids will form in a solution containing dissolved materials similar to those found in primordial seawater when stimulated by light or electricity. These amino acids are not alive, but they are the building blocks of life. To support this hypothesis, we can point to fossils of simple bacteria-like organisms in rocks over 3 billion years old, which are some of the most ancient sedimentary rocks on earth. It is interesting to contemplate that it probably took less than 1 billion years for those simple forms to develop, but it took about 2 billion years to make the step to many-celled organisms.

Life slowly grew ever more complex, branched onto land and continued developing to its present apex on land and in the sea. Some of the resulting ocean creatures defy description: whales over 30 m (100 ft) long, squid of great size and bizarre shape living at almost unimaginable depths, glowing fishes, tiny planktonic creatures on which life on earth ultimately depends for much of its daily oxygen, intelligent porpoises and dolphins, tremendous fisheries that provide food for millions of humans, graceful sea stars that enliven our trips to tidal pools, seaweeds swaying and floating near divers. There are birds with migration routes 40,225 km (25,000 mi) long, turtles that navigate unerringly to a spot near where they were born, and sea lions by the thousands on desolate beaches. There are penguins and pelicans, corals and conchs. There seems to be a little bit of anything alive in the seas.

In the terrestrial sense, the oceans contain and cover all the shapes familiar to us on dry land. There are hidden mountain ranges, submerged peaks higher than any above the waters, hot springs of immense dimension, volcanoes spewing lava in huge underwater eruptions, earthquakes, avalanches, mineral treasures, and more.

The ocean itself is incomparably vast – it contains 361,000,000 cubic kilometers (139,000,000 cubic miles) of water, a mass weighing a staggering 155 billion billion tons! The ocean's volume is eleven times the volume of land above sea level. We tend to divide the ocean into artificial compartments called "oceans" or "seas," but in fact there are no dependable natural divisions, only one great mass of water termed the World Ocean.

In spite of the size dominance of the ocean, the history of the ocean is the history of exploration. Since the beginnings of human development the barriers of ocean have not impeded the movement of humans over the earth's surface. This was aptly illustrated when the European explorers set out to "discover" the world only to meet native populations at nearly every landfall! The tales of adventure and discovery are grand to hear, though the era of exploration is by no means at an end. True, the great land masses have been plotted and traversed, but exploration on and under the seas continues. Perhaps the best is yet to come!

Of course, ocean people today are not always explorers in the traditional geographical sense. Of the nearly 2,500 professional oceanographers working in the United States, about 48 percent are biological oceanographers and 31 percent are physical oceanographers and meteorologists. Many more people work in support of these scientists as marine technicians and technologists and ships' captains and data reducers and computer specialists, but the number of key ocean scientists is not large.

Certainly, the areas to explore and the scientific disciplines involved are many: biology, geology, chemistry, and physics, among others. The text reading assignment for this lesson gives a good overview of the historical development of these disciplines in the study of the oceans. Applying these many fields of study, professionals in oceanography spend their working lives looking for answers to questions that have fascinated humanity since those first days when the Greeks looked past the western reaches of the Mediterranean and said, "Beautiful!" and asked, "Why so?"

Learning Objectives

After completing the reading assignment and viewing the program, the student will be able to:

☐ Recognize the fascination the sea has held for generations of people.

☐ Appreciate the variety of organisms residing in the ocean environment.

☐ Describe some of the features one would see in the underwater world.

☐ Distinguish between the nature of the terms *Atlantic Ocean*, *Pacific Ocean*, or *Indian Ocean*; and the term *World Ocean*.

Key Terms and Phrases

Oceanus The legendary Greek ocean-river.

Ocean The word *ocean* is properly reserved to describe the single great body of water that covers 71 percent of the surface of the badly named Earth. In common use, however, we know of the Pacific Ocean, the Atlantic Ocean, and so on.

Sea One of the larger bodies of salt water. Smaller in size than the ocean. The word *sea* is frequently used interchangeably with ocean.

Technically, and from mythology, there can be many seas but only one ocean.

Evolution The maintenance of life under changing conditions. As conditions change, organisms experience stress. Certain natural variations between organisms permit some organisms to withstand this stress better than others. The organisms that withstand the stress survive and reproduce; those that cannot will die. Some characteristics that allow survival are transmitted to offspring, which then are also capable of survival. By this natural selection the shapes and characteristics of organisms change through time. These changes in shapes and characteristics, when combined with environmental isolation, eventually lead to the appearance of new species.

Species A group of actually (or potentially) interbreeding natural populations reproductively isolated from all other organisms.

Oceanographer An ocean scientist. Oceanographers are by necessity acquainted with the relation of a large number of scientific subdisciplines to the sea. They generally possess graduate degrees in one or more marine science specialty areas such as marine geology, chemistry, marine biology, cartography, or sedimentology. They also usually have broad backgrounds in the fundamentals of science (mathematics, physics, and chemistry), as well as basic skills such as writing. Oceanographers plan experiments and courses of investigation.

Marine technician An ocean paraprofessional, trained in the manipulation of oceanographic equipment and in the analysis and reduction of data. Marine technicians do the work that oceanographers plan.

Before Viewing

☐ Become familiar with the textbook your instructor has selected to accompany the telecourse. This book is an important resource for the course, and it will be most useful if you get a feel for the overall content of the text before you embark on the reading of the individual chapters. Read the preface to the book and look through the table of contents. Thumb through the book and stop at interesting photographs. Read any sections that interest you now. Locate the various Appendixes and note the presence of the helpful Glossary in which you can identify unfamiliar terms. Now you are ready to begin your reading in earnest.

☐ Note the units of measure to be used throughout the book and in this study guide. These are described in the Appendixes in the text.

☐ Read Chapter 1, pages 1–7, and Appendix I in Garrison; or Chapter 1, pages 2–22, and Appendix IV in Ingmanson & Wallace.

☐ Watch Program 1: "The Water Planet."

After Viewing

Reread the text sections, being sure you understand the key terms and phrases, and complete the following exercises:

1. What do you suppose gave the Greeks the impression that the ocean was a flowing river?

2. If this planet is dominated by water, how did it get the name *earth*?

3. Why do biologists largely agree that life arose in the ocean?

4. If oceanography and marine science are so important, why are so few professional oceanographers at work in this country?

Answer Key for Exercises

1. The Canaries Current passes close to the Straits of Gibraltar at the mouth of the Mediterranean Sea. When the Greek navigators ventured past the mouth and into the ocean, they drifted to the south because of this current. Without an understanding of the global nature of the oceans, it would have been a logical assumption that this body of water was really a huge river, impossible to see across, that was flowing to the south. They could not know that the Canaries Current is only one of four huge currents that move water clockwise around the periphery of the North Atlantic Ocean.

2. The word *earth* is derived from the old Anglo Saxon word *eorthe*, which itself is descended from the ancient German word for earth, *Erde*. The Germanic tribes, in their forest fastness, could be forgiven for their lack of understanding of the true nature of the World Ocean. A better name for the earth would surely be *Oceanus*, but so far the authors of this study guide have not had much luck in getting the name changed!

3. The main evidence that life evolved initially in the seas is the presence, within every living thing, of water containing most of the same basic components as seawater. All of cell physiology on this planet seems based on a solution of materials within water. The water provides a fluid matrix within which the processes of life can take place. Because all of the organisms here share this matrix, whether the organisms live on the driest deserts or in the ocean, it is logical to assume evolution from a common origin in the sea.

4. The numbers are deceptive. There may be only around 2,500 professional oceanographers in the United States, but there are many, many more paraprofessionals and business people in association with them. In coastal areas a large number of people derive their income from recreationally related marine industries, from fisheries, from mineral and petroleum companies, and other sources. A Ph.D. and an academic position are not immutable requirements for work within the ocean area. Many universities and community colleges in the coastal areas of the United States have marine science and technology curricula and would be pleased to offer you more information.

Optional Activities

1. Schedule a visit to a research or educational institute and talk to a marine science professional about his or her work. Why did these people decide on this vocation?

2. When you visit the ocean, keep your eyes and ears open. Be a good observer. Look at things and question why they occur. Try to relate what you see in the "real world" to the things you learn through this telecourse.

3. The ocean is an inspiring object and has stimulated a great deal of literature, music, and art. Read some of the poetry of California poet Robinson Jeffers, listen to Ralph Vaughan-Williams's "Sea Symphony," or look at the wonderful abstract seascapes of J. M. W. Turner for a representative sample.

Self-Test

1. Which of the following are generally considered to be within the province of modern marine science?
   a. the deep-diving adaptations of marine mammals
   b. how ocean basins are formed
   c. why earthquakes occur
   d. prospecting for oil
   e. all of the above

2. The ocean covers about _____ percent of the surface of the earth.
   a. 50
   b. 71
   c. 80
   d. 61
   e. 90

3. The concentration of salts and other minerals in the cell fluids of organisms is _____ the concentration of these same materials in the ocean.
   a. very close to
   b. much greater than
   c. much less than
   d. in no way related to
   e. identical to

4. In the scientific method, scientific theories
   a. must be tested and verified by observations
   b. must be verified by the leading authorities in the field
   c. must be consistent with previous, universally accepted scientific concepts
   d. must be consistent with the fact that the ocean is of great age
   e. are accepted as absolute fact until proven otherwise

5. The world ocean
   a. plays a minor role in the weather and shape of landmasses of the earth
   b. does not influence the way organisms live on land
   c. is the dominant feature of the earth and its biological systems
   d. is a common occurrence in the known universe
   e. is a temporary phenomenon on Earth

6. Life on Earth appears to have evolved
   a. in the ocean
   b. on land
   c. in space

Supplemental Reading

Borgese, E. M., *The Drama of the Oceans*. New York: Abrams, 1974.

Broecker, W. S., "The Ocean." *Scientific American*, vol. 249, no. 3 (1983). (This is an excellent overview of the current state of oceanographic understanding.)

Carson, R., *The Sea Around Us*. New York: Houghton Mifflin, 1951.

Dickerson, R. E., "Chemical Evolution and the Origin of Life." *Scientific American*, vol. 239, no. 3 (1978), 70–86.

Hamilton, E., *The Greek Way*. New York: Norton, 1958. (This book is a classic well worth reading.)

Miller, S. L., "Production of Some Organic Compounds Under Possible Primitive Earth Conditions." *Journal of the American Chemical Society*, vol. 77, no. 2351 (1955).

Revelle, R., "The Ocean." *Scientific American*, vol. 221, no. 3 (1969), 54–65.

Slocum, J., *Sailing Alone Around the World* (1899). Reprinted: Sheridan House, 1972. (Astonishing account of the first single-handed circumnavigation.)

Lesson Two   Cosmic Origins

Overview

Earth seems badly misnamed. The very feature that makes the earth unique in our solar system, that makes life possible, that controls the weather and gives the planet its startling blue color from space, is ignored in that name. The ocean is clearly the dominant feature of this lovely and graceful sphere, and, curious though it may seem, these waters were formed through unimaginable violence within a star.

The connection between stars and oceans may seem unlikely, but stars are great nuclear furnaces that achieve the alchemists' goal of transmuting elements – that is, of changing one substance into another. The predominant form of matter in the universe is hydrogen, and stars convert hydrogen to heavier substances, releasing huge quantities of energy in the process. Our earth and its oceans are plainly made of more than just hydrogen gas, so the alchemical machinery of a star is needed to construct the heavy elements from which we are made.

Stars are born when a quantity of hydrogen, dust, gas, and debris coalesces into a tenuous cloud-like sphere, which gradually shrinks under the influence of its own weak gravity. The shrinking continues until compression causes temperatures within the cloud to reach approximately 20,000,000°F, at which point nuclear fusion reactions begin to produce heavy elements at the core.

Such a star may live for billions of years, constructing tremendous quantities of oxygen, carbon, lithium, and other heavier elements from the hydrogen that forms its bulk. When it has converted a certain percentage of its nuclear fuel, the star may become unstable and explode, forcing its shattered mass into space at tremendous speeds. Such an exploding star is called a *nova*. The sighting in February, 1987, of a nova bright enough to be visible to the unaided eye from the Southern Hemisphere has given astronomers an opportunity to observe such an event with sophisticated detectors and telescopes. Data gathered by these devices have confirmed most theories about a star's collapse and the subsequent formation of heavy elements. The earth and its oceans are the indirect result of such an explosion.

Our star, the sun, is believed to be a second-generation star. The condensing cloud that was forming into our sun many billions of years ago was probably struck by the remnant of an earlier exploded star. Our young sun was dramatically influenced in at least two ways by this confrontation. First, spin was imparted to the sun; and second, a fraction of the heavy atom content of the expanding remnant was absorbed into our sun. As a result of these occurrences, planetary material was spun away from the sun as it continued to shrink toward its present size. Some of the resulting planets are gas and vapor bodies. Others, such as Earth and Mars, are rocky spheres that lost their thick early atmosphere when the sun first "turned on." Earth was at the proper place within the scheme of things to retain a large dose of hydrogen hydroxide in its recipe . . . water!

The earliest years of the earth were years of violent change. The earth probably began as a cool, homogeneous body that had collected mass from the material in the vicinity of its orbit around the sun. As time and gravity acted on this sphere, the heat of compression and radioactive elements heated the planet and caused a heavy molten core to form. Lighter substances were forced to the exterior, becoming crust and atmosphere.

The earth's age is generally given as about 4.6 billion (4,600,000,000) years. The point at which the earth clock begins is the time of solidification of the crust – brittle rock that floats, then as now, on a more dense but viscous mantle. When the radiation of heat away from the earth allowed the crustal surface to cool toward the boiling point of water, the stage was set for the greatest rain in the history of our solar system.

The original waters were probably contained in the rocks of the mantle and were released to the surface primarily through volcanic activity. The hot vapors rose and condensed in the cool layers of the upper atmosphere. The rains fell on the young planet only to boil back again into the clouds. This cycle, which lasted for millions of years, accelerated the cooling of the crust and eventually the oceans stabilized in their basins, having taken along an assortment of dissolved minerals leached from the hot solid surface. An alternate hypothesis suggests that a part of the earth's surface water may have come from frozen water-rich comets striking the upper atmosphere.

The physical expanse and distribution of the early oceans is a matter of some controversy. Some researchers have held that huge emergent masses of granite rock (called *sial*) have always protruded through the ocean surface to form continents. However, more recent evidence suggests that the waters of the earth covered the entire surface for some 200 million years before the continents emerged.

Today a large volume, but relatively insignificant percentage, of water remains in the atmosphere as vapor, and some water is within the freshwater lakes and rivers, ice caps, and groundwater mechanisms of the world. However, the vast majority of Earth's water is oceanic; this oceanic water distinguishes the planet Earth from all the other known planets.

Not only does water exist in liquid form on Earth, but ocean water is intimately intertwined with life on this planet: Virtually all forms of life on Earth reflect the composition of the oceans in their body fluids. Though the scientific community cannot agree upon exactly how life evolved, scientists generally agree that:

☐ all forms of life on Earth share the same basic mechanisms for transferring information from one generation to the next (genetic code);

☐ the biochemical machinery for energy conversion and storage is shared by virtually all life forms; and

☐ all life exists, internally, in a fluid matrix very closely resembling seawater. In a sense, all life on Earth is marine.

These ideas have some interesting portents for the origin of life on Earth. For example, it is a fairly safe assumption that life arose in the sea and that the forms of life we see today on the earth probably had a common ancestor in the dim distance of time. The commonality of systems and the presence of oceanlike water in virtually all organisms make this a reasonable conclusion.

Exactly where in the ocean the first forms of life arose is also a matter of debate. In the past it was theorized that the origin occurred on the surface. The rich "earth soup" of the surface (which resulted from the dissolution and concentration of the crustal minerals), the ready availability of all sorts of energizing radiation, and the surface film, which traps small bubbles of air in which active compounds combine and recombine, make the surface a good candidate.

Convincing arguments have been made for life's first glimmer as an abyssal event that occurred in the deep seabed rifts that girdle the earth like seams on a softball. Within these deep, hot spring areas are chemicals leached from superheated rocks and protected from the intense ultraviolet radiation now present at the ocean's surface. It is theorized that these concentrated compounds provided the conditions for cell division and, thus, life on Earth.

Finally, when our star, the sun, becomes unstable and dies or explodes and envelops the earth, this blue jewel world will cease to exist. Its atoms will be distributed into space, perhaps to be incorporated within another planet with another ocean. It is thought that only a small handful of planets in the galaxy have an ocean. There may be lots of rocky planets, and lots of gaseous planets, but a water planet is very, very special. And we named it Earth!

In October 1984 astronomers announced the discovery of what may be a solar system forming around the nearby star Beta Pictoris in the Southern Hemisphere constellation Pictor. This system, discovered by IRAS (infra-red astronomy satellite), was confirmed by photographs taken by the largest optical telescope in the Southern Hemisphere located in Chile. Beta Pictoris is 50 light-years distant (about 300 trillion mi) and may represent an embryo system in a configuration similar to our own about 4.5 billion years ago.

Astronomers have long felt that solar systems are not unusual, and these new pictures are the first direct evidence of planets condensing around another star. Scientists using the Hubble Space Telescope, an optical instrument powerful enough to resolve such systems in our local neighborhood of the Milky Way galaxy, have reported new candidates.

Learning Objectives

After completing the reading assignment and viewing the program, the student will be able to:

☐ Appreciate the uniqueness of a water planet.

☐ Understand, in general terms, the events leading to the formation of the earth and its oceans.

☐ Know the age of the earth.

☐ Compare the major scientific hypotheses for the origin of life on Earth.

Key Terms and Phrases

Transmutation The conversion of one element into another. This may be accomplished naturally in stars or artificially in the nuclear laboratory. It may occur spontaneously during the decay of radioactive elements. Alchemists spent much of their time trying to transmute base metals into gold, but they did not have the nuclear physics to do this successfully.

Star A self-luminous, gaseous, celestial body usually of great size. Generally, stars obtain their energy from the conversion of hydrogen into helium and other heavier elements. The nearest star to us is the sun, some 93,000,000 miles away.

Supernova Popular term used for the period at the end of a massive star's life when it becomes unstable and explodes. The supernova stage is the most dramatic event in a star's life and results in the fabric of the star being flung into space at tremendous speeds.

Planet A subordinate body associated with a star. Our sun has two basic kinds of planets in orbit around it: gas planets (such as Jupiter) and rocky planets (such as Mars or Earth). Planets and stars form from similar processes, but the mass of the star is sufficiently greater to permit nuclear fusion to begin during the compression stage.

Accretion Increase of mass by external accumulation or addition of particles.

Life A system that can capture, store, and transmit energy and that is capable of self-replication (reproduction).

Before Viewing

☐ Read Chapter 1, pages 7–19 in Garrison; or Chapter 2, pages 23–29 in Ingmanson & Wallace.

☐ Watch Program 2: "Cosmic Origins."

After Viewing

Reread the text sections, being sure you understand the key terms and phrases, and complete the following exercises:

1. Why is the earth so special? What conditions would be required to give rise to a planet such as ours with liquid water at the surface?

2. How do scientists know how old the earth is?

3. Would life on other planets, assuming it exists, be like life on Earth?

Answer Key for Exercises

1. For starters, let's look at stars. Most stars visible to us are members of multiple-star systems. If the earth were in orbit around a typical multiple-star system, we would be close to one of the host stars at certain places in our orbit and too far away at others. Some systems have up to nine stars in orbit around each other, and no one knows how many planets are associated with them. Also, not all stars are as stable and steady in energy output as our sun. If we were in orbit around a star that grew hotter and cooler at intervals, our situation would be radically different than it is at the moment. And if we were in a multiple-star system with unstable stars, the whole situation would be even

worse. But our sun is a single star and is relatively stable in heat and light output.

Next, let's look at orbital characteristics. Our earth is in a nearly circular orbit at just the right distance from the sun to allow liquid water to exist over most of the surface through most of the year. That's luck!

Next, consider our "load" of elements. We picked these up during the accretion phase. At our area of orbit there was an unusually large amount of water (or chemical materials that would later lead to the formation of water) and we were again lucky.

So, with a stable star, a pleasant circular orbit that is well placed, and the right raw materials, we are a water planet. This marvelous combination probably is not found in many places in the galaxy.

2. Going to the moon helped. Samples obtained from the surface of the moon have helped to verify information obtained from rock samples, radioactive dating, astronomical evidence such as the age of meteorites, and so on. All of these evidence lines are in agreement that the earth is between 4.5 and 5 billion years old, and 4.6 billion seems like the best estimate we can currently make.

3. Life on other planets would presumably be much different from life here on Earth. Remember, the common denominator for life on this planet is the ocean itself. Life probably arose within it, and all living forms carry an ocean of sorts around within their bodies. If the ocean is removed as a factor from the equation of life, the resulting organisms would, of course, be different.

For example, consider a hypothetical planet without an ocean but with large amounts of liquid ammonia on its surface. If life were to evolve on that planet, the biochemistry of the organisms would almost certainly not include fats or oils because of a chemical incompatibility between ammonia and lipids. Without these fats and oils, membranes are impossible. Without membranes, cells, as we know them, are not possible either. No cells, no earth-type life.

Notwithstanding this argument, life may not be confined to planets with water. Other life forms may exist based upon other "brews." Certainly, though there is a lot of evidence to indicate that the earth is unique

to known planets, the universe is a gigantic place. There are more stars in our local group of galaxies alone than there are grains of sand on an average beach. The thought that we are alone as a planet harboring life within the universe is untenable to many scientists.

Optional Activity

Planetarium shows occasionally have as their theme the origin and nature of the universe. If you are fortunate enough to be near a large planetarium when such a show is scheduled, see it!

Self-Test

(More than one answer may be correct.)

1. Which of these is arranged in correct order, smallest-to-largest?
   a. atom, star, planet, universe, galaxy
   b. atom, planet, star, universe, galaxy
   c. atom, planet, star, galaxy, universe
   d. atom, galaxy, planet, star, universe
   e. atom, universe, galaxy, planet, star

2. Given almost limitless time, the operation of gravity, and large quantities of hydrogen gas, what is a likely first outcome?
   a. A planet will form.
   b. An ocean will form.
   c. A star will form.
   d. All of these will happen simultaneously.
   e. None of the above.

3. Researchers now possess the tools that may help reveal whether Earth is the only planet in the galaxy possessing significant quantities of liquid water on its surface.
   a. true
   b. false

4. The earth is about
   a. 4.6 million years old
   b. 4,600,000 years old
   c. 4,600,000,000 years old
   d. 4.6 billion years old
   e. 460 billion years old

5. The most likely explanation for the origin of the universe involves
   a. a steady state with no beginning and no end
   b. a cataclysmic "bang" and consequent expansion of material
   c. a cataclysmic "bang" and consequent contraction of material
   d. a slow growth through accretion
   e. none of these

6. Our position within the galaxy is
   a. near the center
   b. in a spiral arm, surrounded by dust and gas
   c. above the polar axis
   d. in a special place on the galactic axis, but greatly removed from the main structure
   e. none of these

7. There is now good evidence for planets surrounding other nearby stars. What would you estimate for the probability that at least one of these nearby planets, away from our own solar system, resembles the earth?
   a. The probability is pretty high.
   b. The probability is about 50 percent.
   c. The probability is rather low.
   d. The probability is extremely low.
   e. It is not possible for these planets to resemble the earth.

8. Which of these statements is true for life on Earth?
   a. Life on Earth appears to have arisen in the oceans between 4 and 5 billion years ago.
   b. Living systems on this planet share similar genetic and energy conversion mechanisms.
   c. Life on Earth seems to require an inner environment of seawater-like fluid.
   d. Life can be defined as a system that can capture, store, and transmit energy, and that is capable of reproduction.
   e. All of the above

9. It is inconceivable that other planets could harbor intelligent life.
   a. true
   b. false

Supplemental Reading

Badash, L., "The Age-of-the-Earth Debate." *Scientific American*, Aug. 1989, 90–96. (An excellent summary of the history of the question.)

Hawkings, S., *A Brief History of Time*. New York: Bantam Books, 1988. (A wonderful introduction to a challenging topic.)

Horgan, J., "In the Beginning . . ." *Scientific American*, Feb. 1991, 116–125. (Thoughts on the origin of life by a number of researchers.)

Jastrow, R., *God and the Astronomers*. New York: Norton, 1978.

Jastrow, R., *Red Giants and White Dwarfs*. New York: Harper & Row, 1967. (Although a bit dated, this is perhaps the best and most readable book covering the evolution of stars, planets, and life.)

Maran, S. P., "In Our Backyard, a Star Explodes." *Smithsonian* VR 19 #1; April 1988, 46–57.

Sagan, C., *Cosmos*. New York: Random House, 1980.

Stahler, S. W., "The Early Life of Stars." *Scientific American*, July, 1991, 48–55. (Discusses the turbulent youth of stars.)

Lesson Three   Historical Perspectives

Overview

The history of oceanography is largely the history of exploration. A human being is a curious and restless organism, and even the vast oceans of this planet did not prevent humanity from moving over the earth to settle in its most distant corners. The combination of brain and hand that sets humans apart from other animals led to the adaptation of the simple raft for oceanic travel, and later the construction of boats. These boats and ships were navigated across large and formidable expanses of water by courage, luck, and skill. Between the first appearance of humanity's ancestors some 2 million years ago and the advent of the formal voyages of exploration by Europeans in the late 1400s, virtually all of the inhabitable reaches of the earth were colonized. The powerful contrasting stimuli of curiosity and greed were to force ever more ambitious voyages, and the earliest foundations of what we know today as marine science dealt with ways of making ocean travel more dependable.

The first direct indication of oceanographic activity is through the work of mapmakers (cartographers) in the Phoenician and Minoan civilizations around 800 B.C. These seafaring people had acquired considerable understanding of the Mediterranean and of its tides, currents, and seasonal changes. The early Greeks knew of the ocean beyond the Straits of Gibraltar and considered it to be a vast ocean-river flowing in a huge circle around the island Earth. They called the river Oceanus. The concept of a flowing river may have arisen from the nature of the ocean currents observed just west of the Strait.

A later Greek, Eratosthenes (died 194 B.C.), was the first to calculate the size of the earth. His calculated value was surprisingly close to the actual value. That the earth was spherical was common knowledge among learned people of this time; not until the advent of the European Dark Ages was this crucial idea nearly lost to Western civilization.

Roman conquests brought a new understanding to European geography. By the end of the second century A.D., the limits of Europe were largely defined, Asia and Africa were understood to be much larger than previously reported, and the ocean began to take on its true vast dimension in the Western mind.

Americans have an unfortunate tendency to equate history with Western European history only. Too often it is assumed that no significant advances in oceanographic exploration were made between the fall of Rome and the voyages of Columbus in the 1490s. This assumption is unwarranted, as many crucial discoveries were made during this time. In those centuries before the European Age of Discovery, when thought in Europe had decayed, the Vikings undertook raids to the south and exploration to the west. They had colonized Iceland by 700 A.D., Greenland by 995, and had a stable, functioning colony in North America by about 1000. Their relations with the native Americans were more humane and satisfactory than those of the Portuguese, Italian, and Spanish navigators who followed.

Before the voyages of Columbus, the Arabs developed the compass to aid in navigating over the deserts, and the Chinese constructed elaborate inland waterways to facilitate trade. Most important, the Polynesians ranged over most of the Pacific Basin, making astonishing voyages of exploration and colonization. In all the history of oceanography, no saga is more splendid than that of the Polynesians. These people migrated slowly eastward from southeast Asia across hundreds of stepping-stone islands to the sacred area near what is known today as Tonga and

Samoa. The society was stable there for a long period; then a violent religious war scattered survivors in all directions across the ocean. One group found its way across the largest oceanic space uninterrupted by land to Hawaii, arriving there at about the same time that the Vikings were colonizing North America. Other isolated groups settled Easter Island and New Zealand; some even re-colonized the islands west of the focal points of Tonga, and Samoa. There is speculation that remnant populations may have traveled to South America, the mainland of North America, the Aleutian Islands, and possibly even toward Antarctica!

The awakening of Europe from the Dark Ages demanded sources of financing for all sorts of elaborate and expensive projects. Trade with the wealthy European civilizations was at this time carried on by lengthy and undependable caravan routes from the East. What a boon it would be to trade and finance if contact might be made by sea, by traveling to the west around an earth that might possibly be round! Columbus represented himself to the rulers of Spain as just the man who could make this connection. When he reached islands off the coast of Central America in 1492, he considered himself in India and called the inhabitants he encountered Indians, a name with which native Americans have been stuck ever since. The error in navigation was soon discovered, and suspicions grew that here, indeed, was a fantastically wealthy new world ready to be plucked. Naturally, no consideration was wasted on the fate of the existing inhabitants.

Columbus's four journeys to the area of America (without ever once setting foot in North America) opened the New World to the crushing exploitation of Europe. The Age of Discovery ended with the first circumnavigation by Magellan's crew in 1522.

The first scientific sea voyages were those of Captain James Cook of England in the years 1768–79. This brilliant captain was the first ocean scientist, and his interests and activities were broad and exciting. He made three major trips: The first involved charting the areas of Australia and New Zealand; the second circumnavigated Antarctica about 60° south latitude; and the third led to his death in Hawaii at the hands of outraged native islanders, who incorrectly thought his crew was stealing their chief's boat.

A man of many extraordinary talents, Cook usually displayed consummate diplomacy when negotiating with populations of primitive people who were experiencing their first contact with outsiders. His mandate was to befriend native populations. If territory was needed later for English military bases or victualing stations, negotiations would be less difficult once relationships were established. This skill served him well until his last confrontation with the Hawaiians.

Cook also seemed to have a genuine affinity for exploration. His journals and note-books show an avid interest in the sights and peoples encountered. His powers of observation were excellent; his cartographic skills, exceptional. Among many accomplishments, he proved the use of chronometers in the determination of longitude, established the idea of fruit and vegetable juices to prevent scurvy, and understood the critical necessity of taking scientific samples.

With Captain Cook we reach the beginning of the organized study of the seas.

Following Cook were Charles Darwin in the small ship *Beagle* (1831–36); the polar explorations of Ross, Wilkes, and others (1839–43); and the epic voyage of the steam corvette *Challenger*. The *Challenger* expedition (1872–76) is a landmark because of its superb organization, formal application of sampling techniques, and excellent analysis afforded to specimens collected. The *Challenger* was the first expedition chartered exclusively for the purpose of scientific oceanography (Cook's expeditions were partially military), and the word *oceanography* was even invented by members of this group.

A particularly fascinating chapter in recent oceanographic history deals with the discovery of the principle of plate tectonics, the division of the surface of the earth into vast moving plates discussed in Lesson 9. Early adherents to the theory were hounded as "crackpots," but advances in geology revealed a situation even more bizarre than those first theorists could have imagined. The revolution in earth science is still going on. Plate tectonics, however, is widely accepted and has become a useful working model for understanding the oceans and the continents. The names of Ewing, Hess, and many others have been added to the list of explorers who have changed the way we think about the world.

Your textbook discusses the contributions of many of these "giants" of oceanographic research. As you read this material, try to grasp the essential flow of understanding that has characterized this young science. Notice the gradual progression toward our current understanding of innovative concepts such as abyssal life, continental drift, and current circuits. Modern oceanography moves faster but is fundamentally no different in its approach. Discoveries are still made progressively and by the contributions of a great many individuals.

Because of the huge expense now involved, modern oceanography depends on international consortia and large educational institutions. Private explorations have been replaced by well-orchestrated teams of researchers operating from large and sophisticated ships, from aircraft, and from space. The ocean is recognized as a major planetary resource; international cooperation is more and more critical to a firm understanding of its nature.

One of the most spectacular areas of advancement in oceanographic research through the past five or six years has been, oddly, space. Satellites have made some astonishing discoveries. For example, we have recently found that the ocean surface is actually depressed over the great trenches of the ocean floor, a feature evidently caused by an increase in gravitational attraction in these areas. When the wave activity of the ocean surface is averaged by computers, these gravity-caused depressions become visible on a planetary scale. Satellites can also be used to investigate the primary productivity of the ocean's surface. Masses of phytoplankton, the small drifting plants you will learn about in this course, can be detected from space by their characteristic color-absorbance patterns. Because the density of phytoplankton is directly related to food productivity, these data allow scientists to predict the abundance of food chains and fisheries for that part of the world.

Another exciting and fast-moving area of oceanographic research has to do with computer applications. Powerful "super computers" can do millions of calculations per second. This sort of power is necessary in running programs dealing with the prediction or modeling of weather and ocean currents. Even faster computers will eventually lead to more accurate weather forecasts and an improved understanding of such odd phenomena as El Niño. Also important in oceanographic work is the advent of the "microcomputer revolution," which has brought small computers to offices, research laboratories, and homes at relatively low cost. Quick access to these powerful new tools, and the widespread computer literacy they encourage, has made a great impact in oceanographic research in recent years.

As you will discover in this course, the spirit of exploration still lives. However, today conquering the horizons requires computers, geology, chemistry, molecular biology, and physics.

Learning Objectives

After completing the reading assignment and viewing the program, the student will be able to:

☐ Place in approximate chronological order the major historic contributions to modern oceanography.

☐ List some of the major contributions of Captain James Cook to the science of oceanography.

☐ Identify some of the major contributions of the *Challenger* expedition to the science of oceanography.

☐ Compare the relative importance and difficulty of achievement of the voyages of the Vikings, Columbus, and the Polynesians.

☐ Recognize the characteristics of research and researchers in the marine science field today.

☐ Examine the evolution and progression of maturing theories in oceanography from "crackpot schemes" to accepted scientific constructs.

Key Terms and Phrases

Cartography The art and science of chartmaking. Charts differ from maps: Charts are predominantly of water, maps predominantly of land.

Chronometer A very accurate clock. Curiously, it may not always tell the right time, but the error rate of the instrument is precisely known so that the right time can be derived. The combination of constant error rate and accuracy under very difficult conditions makes a chronometer different from a normal clock. The first marine chronometer was designed in England by John Harrison, a Yorkshire cabinetmaker, in 1728. Through 25 years of painstaking modification and ceaseless effort he perfected the device in time (so to speak) for Cook's voyages. Many wristwatches now sold meet the timekeeping criteria for a chronometer and are quite inexpensive. Far fewer wristwatches, though, meet both the timekeeping and shock-resistance requirements. Most of these are made by the Rolex Watch Company in Geneva, Switzerland, and are still very costly. But just imagine the fun you would have if you could show Captain Cook your $19.00 Casio running watch!

Latitude Imaginary lines on the earth running parallel to one another in an east/west direction. Knowing which line you are on can tell you how far north or south of the equator you are. The equator itself is the zero latitude line.

Longitude Imaginary lines on the earth passing through both poles. Knowing which line you are on can tell you how far east or west of the prime meridian you are. The prime meridian (zero longitude line) runs through the Royal Observatory at Greenwich, England.

The Age of Discovery Begins in 1492 with Columbus's discovery of America and ends in 1522 with the return of Magellan's crew from the first circumnavigation. The term *Age of Discovery* is somewhat artificial because by this time virtually all of the inhabitable land areas of the earth were supporting people. The age does have important sociological and economic consequences for Europe and North America, of course.

Polynesia One of the three principal divisions of Oceania. It comprises those island groups in the Pacific lying east of Melanesia and Micronesia and extending from the Hawaiian Islands south to New Zealand.

Cook James Cook was the first scientific oceanographer. He made three important voyages of exploration and science, and was killed on his last voyage, in Hawaii in 1779.

Challenger The first wholly scientific oceanographic expedition. The name is that of the corvette used in the voyage.

Oceanography Literally, the marking of oceans. The term has come to mean the science of the seas, including their physical, chemical, geological, biological, and geographical components.

Before Viewing

☐ Read Chapter 2, pages 24–53, and Appendixes III and IV in Garrison; or reread Chapter 1, pages 2–22, and Appendixes I, II, and VI in Ingmanson & Wallace.

☐ Watch Program 3: "Historical Perspectives."

After Viewing

Reread the text assignments, being sure you understand the key terms and phrases, and complete the following exercises:

1. Did Columbus discover America?

2. If the Vikings were here first, why don't we speak Danish?

3. How did the Polynesian navigators accomplish their astonishingly long and accurate voyages without modern navigation equipment?

4. What pivotal characteristic made Captain Cook such an excellent man for the job?

5. How does one discover one's position at sea?

6. How can oceanography be conducted from space? Is there really a chance, as the text authors suggest, that oceanographic research can be conducted without ever going to sea?

7. Some scholars suggest that Matthew Maury is the "father of physical oceanography." Others say it is Sir Edmund Halley. Propose an explanation of why this discrepancy exists and explain how both men are important to the field.

Answer Key for Exercises

1. Yes and no. He certainly spent a lot of time investigating islands off the coast, and he did look into bays and harbor areas in Central and South America. But he never actually landed on the mainland of North America. His exaggerated reports of riches triggered an orgy of exploitation, so history gives him credit for the initial discovery.

2. The Vikings accidentally explored the coast of North America in 986 when Bjarni Herjulfsson was blown past his Greenland destination by unfavorable winds. Greenland was a settled Scandinavian outpost. For five days he and his crew sailed the coast of North America, but did not land. His tale awakened a sort of real estate fever in Greenland, and Leif, son of Eric the Red, bought Bjarni's ship and prepared to explore westward. Somewhere in what is today Newfoundland, Leif built a house and spent the winter. He called his discovery Vinland, or Wineland.

   The Greenlanders later mounted a major effort to colonize Vinland. Early in the eleventh century, according to Eric the Red's saga, Thorfinn Karlsefni fitted out three ships, loaded them with at least 160 men and women and adequate livestock, and sailed for North America. The colony lasted for three years. During this time Thorfinn's wife gave birth to the first non-native child born in North America, a boy named Snorri.

   Why did the colony fail? Many hypotheses have been presented. At about this time a climate change resulted in more violent winters on the North Atlantic. The Vikings' open boats would have been difficult to navigate under the best of circumstances in that northern, storm-tossed ocean, but if the weather pattern deteriorated, it could have been impossible to supply the colonists with provisions through the first difficult years. Another explanation has to do with weaponry. Although the Viking colonists tried, with initial success, to co-exist peacefully with the native Americans, eventually violence erupted. The weapons of the Vikings were no better than the weapons of the natives. Columbus succeeded because he had firearms.

   Oh well, we have Danish furniture.

3. The story of Polynesian navigators is fascinating. This priestly art is now nearly lost. The navigators were a special, almost mystic caste with special privileges and powers in Polynesian society. Their secrets were passed from father to son in strict succession. The best navigating families were in great demand.

   Polynesian navigators knew the stars, knew the waves, knew bird species and migration routes, knew wind direction, knew cloud patterns over certain areas, knew the smell and taste of the water near special places, and knew the ocean. They were experts in discerning minute changes in wave patterns and forms, in sunlight color, in places of star-rise and star-set, in the rock of the outrigger canoe, in the way the air felt. They used epic poems and sagas from memory. They befriended the shark and the creatures of the air. They were cousins to Maui, god of the ocean realm. They were magnificent sailors.

4. He was an excellent scientific observer. In addition to the importance of his organizational and administrative skills, this factor made his contributions to ocean science of high quality and dependably accurate. His navigational work was first rate. One could always depend on a chart derived from Cook's positions. In fact, many charts of the South Pacific were Cook-derived right up to World War II! Accuracy and precision made Cook the perfect man for this job.

5. There are many ways. You can now purchase a box to take with you on your ship that reads out, in little lights, the latitude and longitude of your position any time you push the button. But this takes all the fun out of navigation. From the invention of the chronometer (which made longitude calculable) to the early 1960s, the way one routinely determined position was by means of a sextant, a chronometer, and a set of spherical trigonometric tables.

   There has always been a fascination surrounding the seemingly magic mechanism by which a navigator "shoots" the stars, retires to his stateroom, and emerges some minutes later with a scrap of paper bearing the position.

   Latitude (north/south position) is easy to find. Just shoot the angle between the north polar star (Polaris) and the horizon and read the angle from the sextant. That angle is your latitude, approximately. Longitude is trickier. Assume your chronometer, set at English time, reads noon. You look up where you are and find the sun near the horizon almost ready to set. Somewhere over there to the west is England. You measure the angle from straight up (where the sun should be at noon) and where it actually is. This angle tells you how far east of England you are, in degrees. Convert the degrees to miles by knowing the size of the earth.

There's rather more to it than that, but at least you have a sample.

6. Space-based oceanographic research depends on the emission or reflection from the earth of various radiations. Appropriate sensors are placed on satellites, and these radiations are received and analyzed. Some radiations used are visible (visible light); some are invisible (infrared or ultraviolet light). Some are produced by the satellite itself (radar) and beamed down to be reflected off the ocean.

7. Science is a progressive sort of activity. Historians of oceanography usually give credit to Maury for his insightful assembly of data on currents, but Halley was more of a theoretician and innovator. It is hard to say who made more of an impact on the science, so perhaps we'd better not give labels such as "father of oceanography" to anyone!

Optional Activities

1. Try navigating from the stars. You don't need fancy equipment or even a vessel. You can find a star chart in almost any elementary textbook. After identifying a few obvious constellations you can easily find the compass points. Using a protractor you can measure angles between the horizon and bright stars. It is possible to obtain a rough fix even from simple data such as are provided by these methods.

2. Think quietly for five minutes about the difficulties of navigating a small boat across the seemingly endless ocean. Consider the motivation of the Polynesians, the first

Europeans, and the Vikings. Would you go? Whom would you take? What provisions would you require? Why would you go?

3. If you ever visit London, make a trip to the Royal Observatory at Greenwich. Not only will you have the pleasure of straddling the Prime Meridian, but you will have a chance to see Harrison's chronometers in action. The four main clocks have been restored and are functioning perfectly. They are lovely devices, and the sound they make is alone worth the trip!

4. Talk to an oceanographer about his or her job. Does it involve modern explorations?

Self-Test

1. This man is generally credited with the discovery of the nature of the Gulf Stream. He published the first chart of its course.
   a. Matthew Maury
   b. Wyville Thompson
   c. Captain James Cook
   d. Christopher Columbus
   e. Benjamin Franklin

2. This man was the first to comprehend the overall pattern of the ocean currents:
   a. Matthew Maury
   b. Wyville Thompson
   c. Captain James Cook
   d. Christopher Columbus
   e. Benjamin Franklin

3. This man is generally considered the first scientific oceanographer:
   a. Matthew Maury
   b. Wyville Thompson
   c. Captain James Cook
   d. Christopher Columbus
   e. Benjamin Franklin

4. The true size of the earth was first known, within a small range of error,
   a. in the time of Columbus
   b. in the time of Eric the Red
   c. in Greece before the birth of Christ
   d. in the time of Captain Cook
   e. in the time of Benjamin Franklin

5. Polynesian navigators depended on
   _____ for accurate navigation.
   a. luck
   b. stars, clouds, and birds
   c. the appearance and smell of the water
   d. wave direction and shape
   e. all of these things and more

6. The first exclusively scientific oceanographic
   research expedition was
   a. Franklin's researchers from mail ships
      bound for France
   b. Cook's first voyage
   c. Cook's third voyage
   d. the *Challenger* expedition
   e. the *Glomar-Challenger* expedition

7. Which of these contributions lists is arranged in chronological order, earliest-to-last?
   a. Polynesian voyages, Cook, Vikings,
      Columbus, *Challenger*
   b. Polynesian voyages, Vikings, Columbus,
      Challenger, Cook
   c. Polynesian voyages, Columbus, Cook,
      Vikings, *Challenger*
   d. Vikings, Polynesian voyages, Columbus,
      Cook, *Challenger*
   e. Polynesian voyages, Vikings, Columbus,
      Cook, *Challenger*

8. Modern oceanographic expeditions are
   funded, in general, by *all but one* of the
   following:
   a. governments
   b. public foundations
   c. private corporations
   d. private individuals
   e. academic institutions

9. Captain Cook accomplished all of these
   tasks *except*:
   a. discovered the Hawaiian Islands
   b. circumnavigated the world near Antarc-
      tica
   c. made three major voyages of discovery
   d. mapped the coasts of Australia and New
      Zealand
   e. was the first English captain to sail
      around the world

10. Who discovered North America?
    a. Columbus
    b. the native Americans
    c. Drake
    d. Bjarni Herjulfsson
    e. Eratosthenes

11. Satellites use which of the following tools to
    sense ocean conditions?
    a. Doppler radar
    b. photography by visible light
    c. infrared signatures
    d. reflected ultraviolet radiation
    e. all of the above

12. The main force behind the *Challenger*
    expedition was
    a. Osmond Fisher
    b. Charles Piggott
    c. Wyville Thompson
    d. F. B. Taylor
    e. James Hutton

13. The earliest orderly work in what we today
    call oceanography was done by
    a. Alexander the Great
    b. Aristotle
    c. Nick the Greek
    d. Charles I
    e. Tycho Brahe

Supplemental Reading

Bailey, H. S., "The Voyage of the Challenger."
Scientific American, vol. 188, no. 5 (1953), 88–
94.

Bellwood, P. S., "The Peopling of the Pacific."
Scientific American, vol. 243, no. 5 (1980), 174.

Bellwood, P. S., "The Austronesian Dispersal
and the Origin of Languages." *Scientific
American*, July 1991, 88–93. (The Polynesian
voyages are reflected in their languages.)

Elachi, C., "Radar Images of the Earth from
Space." *Scientific American*, vol. 247, no. 6
(1982), 54.

Hough, R., *Captain James Cook*. New York: W.W.
Norton, 1994.

Kane, H., *Voyage: The Discovery of Hawaii*.
Honolulu: Island Heritage, 1976.

Kerr, R., "The Deepest Hole in the World."
Science, vol. 224, no. 4656 (1984), 1420. (This
is a remarkable account of a new exploration
recently undertaken to drill into the earth's
mantle.)

LaFey, H., *The Vikings*. Washington, D.C.: National Geographic Society, 1972.

Landstrøm, B., *Sailing Ships*. New York: Doubleday, 1978.

Morison, S. E., *Admiral of the Ocean Sea*. New York: Oxford University Press, 1942.

Morison, S. E., *The Great Explorers*. New York: Oxford University Press, 1978.

Olson, S., "The Contours Below." *Science 83*, vol. 4, no. 6 (1983).

Williams, F., and Matthew Maury, *Scientist of the Sea*. New Jersey: Rutgers University Press, 1963.

Lesson Four   The Waters of the Earth

Overview

Water is such an abundant, commonplace substance on this planet that we tend to overlook its really remarkable physical properties. For example, we believe that life originated in a water environment, possibly near hydrothermal springs on the seafloor, and was probably confined to those ancient seas for the greater part of earth history. Also, as terrestrial vertebrates, we carry our seawater heritage within our saline body fluids. Even our visual perception of the world is based on what we see through a salty fluid, the aqueous humor that partly fills our eyes.

What are some of the special characteristics of water? (What an interesting situation! Here we are, mostly water, using our brains, which are mostly water, to investigate water!) Probably the most important property in terms of life is the ability of water to dissolve most substances and hold great quantities in solution. It is truly a universal solvent, a complex broth of dissolved organic and inorganic compounds, which provides the nourishing environment for the myriad of organisms living in the sea.

In addition to being a solvent, water has other interesting properties. For example, it has a strong tendency to stick to itself, unlike gases, which disperse rapidly. This tendency, called *cohesion*, combined with *adhesion* (a tendency to stick to other materials) results in a high surface tension. Water acts as if it has a skin on top. A needle or razor blade will float on this skin, and water striders, unusual aquatic insects, walk about on it, only slightly denting the surface with their light weight.

Water is not perfectly transparent, but it does transmit sufficient sunlight for marine plants to photosynthesize in the shallow depths of the sea.

Water also transmits sound more efficiently than air, a fact that enables the great whales to signal to each other over surprisingly vast distances within the sea. The clicks and squeals of the echo-locating porpoises and the grunts and croaks of many fish and invertebrates make a veritable underwater cacophony.

Water is also incompressible and cannot be squeezed into a small space. Although the pressures encountered in the deepest trenches are enormous, up to 15,000 lb/in.^{2}, the water there is still liquid. The strange life forms that exist in the cold fluid of the deep sea go about their activities unhampered.

Water has a high heat capacity, higher than most common solids and liquids, meaning that it takes a great amount of heat energy to raise the temperature of water. Not only is it slow to heat, it is also slow to cool. Ocean temperatures do not vary greatly, allowing coastal cities to enjoy the typical marine climate: cool in summer and warm in winter. Inland regions are not so fortunate.

The density of water (mass per unit of volume) is an important characteristic and is influenced by temperature, pressure, and, in the case of seawater, salinity. Like most liquids, water expands and becomes less dense as it is heated. As it cools, its density increases as the molecules pack closer together. Pure water reaches a point of maximum density, 1.00 g/ml, at 4°C while still in the liquid state. Further cooling causes a surprising decrease in density until at 0°C it solidifies into ice at a density less than that of liquid water. Water is one of the few substances known in which the solid state will float in its own liquid. If ice did not float, consider how different the earth would be. The polar seas and the deep world ocean might be a mass of ice from the bottom up. Lakes and ponds would freeze solid in winter and some would never thaw completely. If ice were denser than liquid water, life might not have developed on this planet.

Water has a relatively high density for a naturally occurring liquid. Oil, with a much higher viscosity, floats on water, as evidenced by oil spills and salad dressings. Density is an especially important characteristic in the study of ocean waters. As noted earlier, the density of pure water is 1 g/ml at 4°C. In contrast, the density of seawater is slightly more, ranging from 1.020 to 1.030 g/ml, depending on temperature and salinity. While density in the oceans increases with depth, the increase is not always uniform. A particle of surface water that has an increased density either from chilling or from an increase in salinity due to evaporation will sink until it reaches a level of equal density. It will then move horizontally along the appropriate density layer. Because density increases with depth, the oceans generally do not overturn, causing little vertical mixing. Only when less dense water occurs below water of higher density will the water column become unstable, overturn, and mix until a new equilibrium is established.

Many of water's unusual properties – its solvent characteristics, high surface tension, and high heat capacity – are related to the structure of water molecules. Water itself is a compound composed of two gases, oxygen and hydrogen, expressed in the familiar formula H_2O: two atoms of hydrogen and one atom of oxygen. The sharing of electrons between these atoms creates an extremely stable compound. In fact, the oxygen is so tightly bonded to the hydrogen that marine organisms are unable to break the water molecule and utilize the essential oxygen. Instead they must depend on free, uncombined oxygen molecules dissolved in the water for their respiration.

Seawater is not too different from pure water in its properties; it is 96.5 percent water and 3.5 percent dissolved salts. The total amount of dissolved salts in the ocean is called the *salinity* and is expressed by oceanographers at 35‰ (parts per thousand). This means that in every 1000 kg of seawater there are about 35 kg of dissolved inorganic substances. "Salts," however, does not refer to just table salt, sodium chloride; it includes many elements. Of these, less than a dozen account for approximately 99 percent of all constituents in seawater, with chlorine being the most abundant (over 55 percent by weight). These substances have been contributed to the oceans over the eons by complex chemical, physical, and biological processes, and then concentrated by evaporation.

The dissolved salts account for some of the properties of seawater that differ from those of fresh water. For example, pure water will freeze at 0°C, but the freezing point of seawater depends on the salinity, with saltier water having a lower freezing point. Seawater, with a salinity of about 35‰, must be cooled to almost –2°C before it starts to freeze. The fishes and other cold-blooded animals in the polar seas live in waters that are below freezing, and although their body fluids are not as salty as seawater, they do not freeze! (They manufacture their own antifreeze!) When these polar seas do freeze and form sea ice, the resulting ice crystals tend to reject the dissolved salts and the ice will have a lower salinity than the surrounding water. Many schemes have been proposed to tow icebergs to areas with water shortages, but so far none has been attempted.

Pure water does not conduct electricity, but the ions in seawater readily conduct electrical current. This property is used to determine the salinity of seawater samples.

All these special properties of water have allowed life to develop and thrive on this single planet. If Earth were not just the right size and temperature to hold water in its liquid state, this planet would have the same barren landscape as the moon or Mars. To a visitor from outer space, Earth without water would be like the rest of the solar system, interesting to visit but certainly no place to live.

Learning Objectives

After completing the reading assignment and viewing the program, the student will be able to:

☐ Describe the remarkable properties of water and understand their significance.

☐ Relate the structure of the H_2O molecule to the special properties of water.

☐ Name the major ions dissolved in seawater and understand the concept of residence time.

☐ Understand the physical conditions of the sea, in terms of salinity, temperature, density, and pressure.

☐ Relate the distribution of dissolved oxygen and carbon dioxide in the sea to biologic and physical processes.

☐ List and describe water-sampling instruments and techniques used on modern research vessels.

Key Terms and Phrases

Hydrogen bonds The attraction of the positively charged hydrogen of one water molecule for the negatively charged oxygen of another water molecule. Hydrogen bonds account for some of the special properties of water, such as high surface tension, high boiling and freezing points, and high heat capacity.

Salinity The total quantity of dissolved solids in seawater per parts per thousand by weight; the number of grams of solids contained in one kilogram of seawater. Standard Sea Water is defined as 35‰ (parts per thousand) at 0°C. The range of salinities in the ocean lies between 34‰ and 37‰.

Law of constancy of proportion The ratio of the major elements dissolved in seawater to each other remains constant, although the total salinity may vary slightly in different places and with depth.

Three-layered ocean The temperature structure of the ocean, consisting of three zones: the mixed layer, the main thermocline, and the deep water.

Mixed layer The warm surface waters of the ocean down to about 200 m. The least dense water of the ocean, it is well mixed by waves and convection currents, and the temperature is about the same top to bottom (isothermal). It is affected by seasonal changes. The mixed layer makes up about 2 percent of the total ocean.

Main thermocline The zone lying between about 200 and about 1,000 m, in which the temperature drops rapidly with increasing depth.

Deep layer The cold, dense waters, including polar waters, that make up about 80 percent of the world oceans. The deep waters are generally unaffected by seasons, by surface currents, or by other surface phenomena. The temperature lies between 0°C and 5°C.

Conservative properties Those constituents of seawater that change very slowly in long geologic cycles, such as salinity, temperature, density, and inert gases dissolved in seawater.

Nonconservative properties Those constituents of seawater that are tied to biologic cycles and may change rapidly in short seasonal or daily cycles. Dissolved oxygen, carbon dioxide, and nutrients such as phosphates, nitrates, and silicates are nonconservative.

Dissolved gases In addition to salts, various gases are dissolved in seawater. Oxygen and carbon dioxide are the most important because of their relation to biologic activities in the sea. The amount of gas seawater will hold in solution depends on temperature, salinity, and pressure. Cold water, less salty water, and water under pressure will all retain more dissolved gas. Polar seas, for example, are high in dissolved oxygen because of the very low temperatures.

Dissolved oxygen Oxygen (O_2) is needed by most animals, including those in the sea, for respiration. Marine animals remove the oxygen dissolved in the water (not the oxygen in the water molecule) by means of gills or other respiratory structures. Oxygen enters the sea from the atmosphere, or as a by-product of photosynthesis, both surface phenomena. The amount of dissolved oxygen in each liter of seawater will vary, from almost none (anoxic conditions) to about 8 ml/l. The highest concentrations will be in the upper 20 m or so of the sea, in the photic zone where plant plankton are actively photosynthesizing. The amount of oxygen decreases downward, reaching a minimum at about 700 to 1,000 m in a zone of sluggish water. There is usually a small increase toward the bottom, due to cold, oxygen-laden currents coming in from polar regions.

Carbon dioxide Carbon dioxide (CO_2) is released during the process of respiration and is removed from seawater during photosynthesis. The amount of carbon dioxide in the sea increases with depth because plants live only near the surface in the photic zone, but animals give off CO_2 at all depths to the very bottom of the seas. Carbon dioxide is very soluble in water, and the sea can absorb tremendous amounts from the atmosphere. In spite of the extensive burning of fossil fuels, which releases CO_2 into the air, the amount in the atmosphere has remained small, about 0.03 percent by volume. There is a slight yearly increase; the rest is apparently going into the ocean.

Density The ratio of the mass of any substance to its volume. Pure water reaches its maximum density of 1.000 g/ml at 4°C. The density of seawater varies between 1.020 and 1.030 g/ml, increasing with increased salinity and pressure and decreased temperature. For seawater with salinities greater than 24.7‰, which would include most of the oceans, the density increases as the temperature is lowered until the water starts to freeze. There is no single temperature of maximum density of seawater, as the freezing point depends on the salinity. Slight differences in density can cause vertical movements of water in the ocean. The density in the sea generally increases with depth, preventing wide overturn of the waters.

Residence time The average length of time an atom or a particle of an element stays in the ocean before removal by precipitation into the sediments or uptake by organisms. The salts in the sea seem to be in a geochemical equilibrium, with the entering salts balanced by the salts removed. The water itself is being recycled as part of the global pattern of evaporation and precipitation, with a residence time in the sea of about 54,000 years.

Latent heat The quantity of heat lost or gained by a substance undergoing a change in state.

Before Viewing

☐ Read Chapter 6, pages 144–165 and Chapter 7, pages 169–182 in Garrison; or Chapter 7, pages 95–108, and Chapter 8, pages 109–123 in Ingmanson & Wallace. Note carefully the figures and tables in the chapters of either book. All are important in understanding the physical and chemical properties of seawater.

☐ Use the glossary in your text to look up the following terms and expressions:

bathythermograph	hydrogen bond
bottom water	Nansen bottle
chlorinity	oxygen minimum
colligative property	layer
compensation depth	pycnocline
condensation	salinity
diffusion	salinometer
freezing point	salt
heat capacity	surface tension

☐ Watch Program 4: "Waters of the Earth."

After Viewing

Reread the text sections, being sure you understand the key terms and phrases, and complete the following exercises:

1. List the special properties of water that are related to the polar molecule and the hydrogen bonds.

2. Describe all the natural processes that might *increase* the surface salinity and the processes that might *decrease* the surface salinity in any one locality. Where are the surface salinities of the world ocean highest? Lowest? Why? In fact, why is the ocean salty when the rain and streams that contribute water to the seas are fresh?

3. List the methods by which oxygen enters seawater and the ways in which it may be removed. What factors might contribute to the oxygen minimum zones? How do the deep water organisms, far from the sea surface, get their oxygen?

4. By what methods does carbon dioxide enter and leave seawater? What is the significance of the compensation depth?

5. Describe the conditions in each layer of the three-layered ocean. Why is there no good thermocline in polar waters? To what depth do seasonal changes affect the temperature structure of the open sea?

6. Consider a small, soft-bodied sea cucumber living at depths of 11 km. About what is its *internal* pressure? What is your internal pressure? If pressure decreases downward in the sea at the rate of 1 atmosphere per 10 m, what is the pressure in the air bladder of a rockfish swimming at 1,000 m? What would this be in pounds per square inch? Why doesn't the rockfish explode?

7. Why are two thermometers (protected and unprotected) placed on the Nansen bottles during a hydrocast? Which thermometer will have the higher reading? If you are using Ingmanson & Wallace, see pages 106–108 in the text. What is the most efficient method at present for measuring sea surface temperatures?

Optional Activities

1. Dissolve about two or three tablespoons of table salt in a glass of water. Add a few drops of food color (optional). Fill an ice cube tray with the salt solution and place in the freezing compartment of your refrigerator.

   Check every hour until the salt water freezes and describe the results. Compare the ice cubes with those of fresh water. The properties of a solution that change with the addition of salts are called *colligative* properties and include boiling and freezing points. What property of the water was changed by the addition of salt?

2. Remove one of the colored salty ice cubes and drop it into a glass of water at room temperature. What happens? Drop in a freshwater ice cube. What does this experiment tell us about the effect of temperature and salinity on the density of water? Let the glass stand for several hours and observe the rates of diffusion or mixing.

3. Check Figures 6.14 and 6.15 on page 157 of Garrison, or Table 7.7 on page 103 of Ingmanson & Wallace. In Garrison, note the areas of high and low sea surface temperatures in Figure 6.14, and the areas of high and low sea surface salinities in Figure 6.15. Can you explain the distribution of these areas? In Ingmanson & Wallace, note the salinity, evaporation, and precipitation of the latitudes 20°N to 20°S. Where is the salinity the lowest? Note the E and P columns and explain the anomaly at 5°N.

Self-Test

1. The hydrogen atoms in a water molecule tend to bond strongly to
   a. each other
   b. oxygen atoms of another water molecule
   c. hydrogen atoms of another water molecule
   d. all positively charged ions
   e. oil droplets in the water

2. The hydrogen bonds of water molecules account for which of the following?
   a. Water is the universal solvent.
   b. Water has a high surface tension.
   c. Water has a high heat capacity.
   d. Water has a high boiling point.
   e. All of these are relevant.

3. The average salinity of the world ocean is closest to which of the following?
   a. 3.5‰
   b. 21.5‰
   c. 35‰
   d. 52‰
   e. 96.5‰

4. The two most abundant elements dissolved in seawater are
   a. fluorine and iodine
   b. gold and silver
   c. bromine and boron
   d. sodium and chlorine
   e. carbonate and sulfate

5. Local surface salinity is increased primarily through
   a. precipitation of rain or snow
   b. freezing
   c. melting ice
   d. evaporation
   e. precipitation of substances out of a solution

6. The saltiest open-ocean water is
   a. in the North Atlantic subtropical zones
   b. in the Pacific equatorial zone
   c. in the Arctic surface water
   d. off Finland
   e. in the Antarctic surface waters

7. The major biologic source of dissolved oxygen in the sea is from
   a. whales and porpoises
   b. fish respiration
   c. bacterial decay
   d. photosynthesis by plant plankton
   e. all of these

8. The oxygen content of seawater, particularly in the Pacific,
   a. is constant top to bottom
   b. decreases evenly with depth
   c. increases evenly with depth
   d. is too small to be measured accurately
   e. is at a minimum at about 700–1,000 m

9. The amount of gases seawater can retain in solution will increase as
   a. the temperature decreases
   b. the salinity decreases
   c. the pressure increases
   d. all of these
   e. none of these

10. The average length of time an element spends in the ocean is known as its
    a. thermocline.
    b. residence time.
    c. calligative property.
    d. mixing time.
    e. solution time.

11. Which of the following zones in the open sea is isothermal and has the warmest and least dense water?
    a. main thermocline
    b. mixed layer
    c. equatorial bottom water
    d. deep and bottom waters
    e. polar surface waters

12. Density of seawater is least affected by which of the following?
    a. temperature
    b. salinity
    c. hydrogen ions
    d. depth
    e. pressure

Supplemental Reading

Broecker, W. S., "The Ocean." *Scientific American*, vol. 249, no. 3 (1983), 146–60.

Lesson Five   Ocean's Edge

Overview

When most of us think *ocean*, we usually visualize the shore or beach, for the deepest part of the sea is still an almost unknown landscape, visited by fewer people than have set foot on the moon.

For the oceanographer, the shore is a laboratory, an accessible part of the ocean where the dramatic interaction of forces that shape the coast can be observed and analyzed. The waves, tides, and currents from the sea; the rivers and glaciers from the land; daily weather changes; and seasonal cycles and long-term climatic variation all play their part in the changing coastal scene.

Of all the processes active in the shore zone, the waves are by far the most significant. Bit by bit, grinding and abrading away the land, the endless parade of waves cuts cliffs, erodes sea caves and sea arches, builds sand bars and spits, and creates the landforms characteristic of the shore. The cliffs retreat under the onslaught and the shoreline creeps landward. We wonder why the continents weren't planed off to sea level millions of years ago.

Sometimes the rocks at the shore are very hard igneous rocks. They resist wave attack and jut seaward as points of land, to the dismay of the unwary mariner. If the rocks are soluble, the waves and spray will dissolve out portions, leaving intricate patterns resembling abstract sculpture. Even the grains of sand brought to the beach by the rivers are tumbled and sorted by the waves and finally washed into deeper waters to become part of the sediments that carpet the seafloor.

Another factor of interest to the oceanographer is the cyclic changes taking place at the shore. The most casual one-day beach visitor can't help but be aware of the ebb and flow of the tides. Twice daily the waters rise on most beaches and drain away, exposing the lower intertidal zone. For the plants and animals that inhabit this variable environment, low tide is a perilous time of hot sun, drying air, and browsing predators. The returning high water brings cooling relief, food, and oxygen, but also the new stress of wave shock. In spite of the dangers and drastic changes, the intertidal zone is teeming with life. It is the most productive area of the ocean.

These tidal changes are short-term cycles that affect only a small portion of the coastal zone. Longer seasonal or yearly cycles may affect the whole beach. In winter large waves come in to the shore from the fierce storms at sea. These crashing breakers strongly erode the beach face, carrying away the sand and exposing the rock platform beneath. Fortunately, in the summer the small waves bring back the sandy mantle just in time for the vacationing beach enthusiasts.

Still longer cycles may affect the margins of the continents throughout the world. For example, during the Pleistocene Ice Age of the last 1–2 million years, ice sheets expanded on the continents removing water from the oceans. The sea level dropped 350–400 ft, and the shoreline retreated to the outer edge of the continental shelf. Waves were breaking as much as 100 mi off the present shore. Mammoths and other Ice Age animals roamed these newly exposed lands and were probably hunted by early humans. Four times the seas retreated and came back, responding to the expansion and melting of the ice caps.

The present sea level is a geologically youthful feature, a mere 6,000–7,000 years old. The last great rise of sea level during the past 15,000 years or so must have been noted with considerable apprehension by early humans and may be the source of Great Flood legends. In the past 100 years, sea level has risen more than one foot (30 cm) and may rise as much as another three feet in the next century. If the present warming trend continues, and the ice caps on Antarctica and Greenland melt, what might happen to our coastal urban communities?

The greatest cycles of land and sea must be measured in millions of years, for they are related to the great movements of the earth's crust. We know that the oceans have changed size, shape, and location over geologic time. If we could somehow project the shoreline on the world globe and observe the changes in position since the beginnings of earth time, we would see it flicker over the landmasses, as the seas repeatedly flooded almost whole continents.

The waterways of the past can be charted from the exposure of thick marine sedimentary deposits found on the interior of the continents of today. The most ancient sedimentary layers, deposited in the first oceans, have been almost obliterated by the erosion, deformation, and recycling of the earth's crust during the past 4 billion years. We get only tantalizing glimpses, in small scattered areas on land, of the bizarre, soft-bodied life forms that populated those primeval waters – faint clues to our own origins. The record improves, however, at the beginning of the Paleozoic era about 600 million years ago. The seas were widespread and life was abundant. The fossils found in these marine strata reveal the remarkable sequence of life forms leading to the populations of the present seas.

The seas today are restricted, but we have every reason to believe the shoreline will continue to fluctuate as it always has in the past.

Humans, too, interact with the shore. We build in this dynamic environment and are appalled when our structures are carried away by waves and storms. We sprawl over the precious wetland and marshes that are the nurseries for the young marine animals, then deplore the decline in marine life. We dump our wastes into the coastal zone and lament the mercury and lead in our food fish.

We must strive to understand the forces and cycles that create the shore zone. Only then will we be able to live in productive harmony with land and sea.

Learning Objectives

After completing the reading assignment and viewing the program, the student will be able to:

☐ Recognize the shore as a changing, dynamic environment.

☐ Relate the changes that occur on the beach to wave energy.

☐ Understand both the short-term and long-term cycles that affect the changing shoreline.

☐ Classify the various kinds of coastlines of the world by origin.

☐ Name the most abundant minerals found in both continental and island beach sands and discuss their origin.

☐ Evaluate the effects of human beings on the fragile coastal environment.

Key Terms and Phrases

Shoreline The boundary between sea and land, usually measured at the high-tide mark or the highest reach of the winter waves.

Beach The zone of unconsolidated sediments (sand, pebbles, shingles, sometimes boulders) and active sediment transport. The landward limit may be vegetation, a sea cliff, sand dunes; the seaward limit may be the lowest low-tide mark.

Wave erosion The result of wave energy acting to abrade the shore, dissolve the more soluble minerals in the rock, or force open small cracks in the rocks by hydraulic action. Coastal features associated with wave erosion include sea cliffs, sea stacks, sea caves and arches, wave-cut benches and terraces, steep rocky beaches. The final result of wave erosion is to cause the land to retreat.

Wave deposition The transport and deposition of sediment on the shore as a result of wave energy. Beach sands, the littoral transport of sand parallel to the beach face, the formation of sand bars, spits, barrier beaches, and the filling of harbors and lagoons are all associated with wave deposition. Deposition may result in the land being built seaward.

Sand Unconsolidated beach sediment, ranging in size from about 0.5 mm to about 2 mm. Sands on any one beach are remarkably well sorted by size by wave energy. The finer particles will be carried out to sea, and the larger grains tend to be left upstream. The remaining materials left on the beach will fall within a narrow range of sizes.

Continental beach sand Sediments derived from weathering and erosion of continental rocks, such as granite, which are composed primarily of quartz, a clear glassy mineral that is very hard and resistant to chemical change. Feldspar, also derived from granite, is another common mineral in beach sands.

Tropical or island beach sands The famous black sands of Hawaii, for example, consist of broken particles of basalt lava, the volcanic material that makes up oceanic islands. These sands include particles of black volcanic glass (obsidian), cinders, basalt, scoria, pumice, and basalt pebbles.

The green sands are also volcanic in origin and include grains of the clear green mineral olivine.

Sands in tropical regions may be white, composed of calcium carbonate fragments eroded by the waves from offshore coral or carbonate reefs. These sands are made of bits of coral, algae platelets, spines of sea urchins, shells of many mollusks, and hard parts of many other sea organisms.

High-energy beach That part of the coast that is subject to strong wave erosion. The high- energy beach will be steep, narrow, and usually covered with pebbles or boulders. The smaller sand grains will have been removed by wave action.

Low-energy beach The broad, gently sloping beach affected by small waves only. It is covered by fine sand-sized particles.

Summer-winter beach regime Certain beaches, because of their location and the direction they face, receive different wave energies during the seasonal cycle. In winter large storm waves remove the sand from the beach and expose the rocks below. In summer small waves replace the sand on the beach.

Longshore current A flow of water parallel to the shore within the surf zone that is driven by waves coming in at an angle to the beach face. Longshore currents result in movement of sand on the beach called *littoral drift* or *longshore transport*. Beach sand is always in transit from one place to another. In the United States the longshore current moves sand toward the south on both the East and West Coasts, as most of the waves that generate the flow of water come from storm areas of the North Atlantic and the North Pacific.

Before Viewing

☐ Read Chapter 12, "Coasts," in Garrison; or Chapter 13, "Coastal Processes and Estuaries," in Ingmanson & Wallace. Using Table 12.1 on page 299 in Garrison, or Table 13.3 in Ingmanson & Wallace, compare the size of beach sediments with the average slope of the beach face. Make a general statement regarding this relationship.

☐ Use the Glossary to define the following terms:

bar	littoral zone
barrier beach	longshore current
beach	low-energy
berm	environment
high-energy	refraction
environment	spit
hook	

☐ Watch Program 5: "Ocean's Edge."

After Viewing

Reread the text sections, being sure you understand the key terms and phrases, and complete the following exercises:

1. Draw a cross section of a low-energy beach, bounded by sand dunes on the landward side and by a shallow longshore bar on the seaward side. Label the backshore, the foreshore, the inshore, and the offshore. Where will the waves be breaking at low tide? See Figure 12.15 on page 300 in Garrison, or Figure 13.1 in Ingmanson & Wallace.

2. List the two major sources of energy that affect the coastline. Which features along a coast result from *wave erosion* and which are from *wave deposition*?

3. What is the major problem threatening the sandy beaches of much of the United States? State the problem, the natural causes, and the causes of human origin. What is being done to alleviate the problem?

4. Describe the shorelines of the world during the maximum advances of the glaciers during the last Ice Age. Where were the beaches located? Where were the waves breaking? What was the climate at that time? Who was living 1 million to 10,000 years ago? What happened to the shoreline when the ice began to melt?

5. What is the general basis for the classification of coasts devised by Shepard? See the discussion in Garrison on pages 290–291, or see Table 13.6 on page 248 in Ingmanson & Wallace. What geologic agent is responsible for the formation of fjords, drumlins, and moraines? Where are you most likely to encounter coral reef coasts?

6. Describe the origin of Long Island and Cape Cod. Can you explain why similar features are not found in southern California? (*Clue*: Did Ice Age glaciation reach southern California? There were mountain glaciers, but the great ice sheets did not reach California.)

Optional Activities

1. Start a collection of beach sands. There is infinite variety in color, particle size, mineral content, rounding, and sorting of sand grains. An inexpensive hand lens, magnifying glass, or low-power microscope will reveal the gemlike quality of the colorful grains.

2. Think of your last beach trip. Was it to a high- or low-energy beach? Summer or winter beach? Shaped primarily by marine or nonmarine agencies? Strong longshore current? Rip currents? Fine or coarse particles on beach? Analyzing beaches is fascinating and makes one aware of the processes at work. Watch for manmade structures that interrupt the beach regime. Try to anticipate the long-range effects of these installations.

Self-Test

1. The most important form of energy shaping the coast is
   a. seafloor currents
   b. wind
   c. waves
   d. longshore currents
   e. tides

2. The energy that drives the longshore currents comes from
   a. tides at full and new moon
   b. waves breaking at an angle to the beach face
   c. winds from inland deserts
   d. currents, such as the Gulf Stream
   e. convection currents in the mixed layer

3. Of the following, which minerals are most likely to be found on continental beaches?
   a. olivine and obsidian
   b. diamonds and gold
   c. manganese and salt
   d. calcium carbonate and gypsum
   e. quartz and feldspar

4. If a beach is wide, gently sloping with fine sands, we would expect to see
   a. very heavy breakers
   b. a beach facing into Arctic storms
   c. generally small waves summer and winter
   d. high-energy waves all year round
   e. wave erosion as the dominant process

5. The famous fjords of Norway originated as
   a. wave-cut benches
   b. deeply eroded glacial valleys
   c. drowned rivers
   d. volcanic explosions
   e. coral lagoons

6. The steep cliffs and rugged coast of much of the West Coast of the United States are primarily a result of
   a. biologic activity
   b. marine deposition
   c. river deposition
   d. glacial erosion
   e. faulting and earth movements

7. The New England coast of Rhode Island and Massachusetts has been modified primarily by
   a. faulting and folding
   b. mangrove swamps
   c. presence of hard limestones
   d. glacial erosion and deposition during the Ice Age
   e. formation of river deltas

8. The development of large sand spits and hooks, such as Cape Cod, indicates
   a. limited sediment sources
   b. offshore coral reefs
   c. transport by longshore currents
   d. erosion by the Gulf Stream
   e. limited wave energy

9. The loss of sand along the southern California beaches is related to
   a. arid region with few large streams
   b. damming of the streams
   c. longshore current carrying sediment to submarine canyons
   d. deposition in harbors and marinas
   e. all of these

10. The beach features seen today on the coasts of the world
    a. are ancient, dating back to the origins of the oceans
    b. are permanent landforms
    c. are stable structures in equilibrium between land and sea
    d. are dynamic, geologically youthful, existing only since the end of the Ice Age
    e. none of these

11. All of the following coastline types *except* _____ would be classified as primary coasts.
    a. fjords
    b. lava flow coasts
    c. deltas
    d. barrier islands
    e. faulted coasts

12. All of the following devices *except* _____ are used to prevent beach erosion.
    a. seawalls and bulkheads
    b. dams on rivers
    c. jetties and breakwaters
    d. vegetation
    e. beach nourishment

Lesson Six   The Intertidal Zone

Overview

Anyone who spends time at the shore, especially a rocky shore, is soon struck by a curious paradox. The shore looks like an extremely difficult place for organisms to make a living, yet it abounds with life. Indeed, the intertidal zone is one of the earth's most densely populated areas. The tide rises and falls, alternately drenching and drying out the animals and plants. Predators and grazers from the ocean visit the area at high tide, and those from the land have access at low tide. Temperatures can change 50° or 60°F in seconds when cold waves crash upon sunlit shells. The powerful shock of the waves tears at the underpinnings and structures of animals and plants. The substratum on which the organisms exist may be unstable. Yet, incredibly, the richness and diversity in the intertidal zone are unmatched in all of the ocean. Thousands of kinds of living things succeed in making their homes in this rigorous area.

The most obvious and important variable in this life zone is the rise and fall of the tides. Virtually all of the coastline of the continental United States experiences two high tides and two low tides each day. For reasons we will discuss in Lesson 14, these tides vary in height and number through time. The variation of tides directly influences the number of hours each intertidal organism must spend exposed to the air. These hours of exposure have a strong influence on the success of organisms.

The animals and plants living on the shore experience conditions much different from those residing at a lower level. It is a simple matter to chart the height of a spot above the Tidal Datum (which, for the West Coast of the United States, is mean lower low water) versus the number of hours that spot is exposed to air, as you see in Figure 6.1. Because some organisms are able to

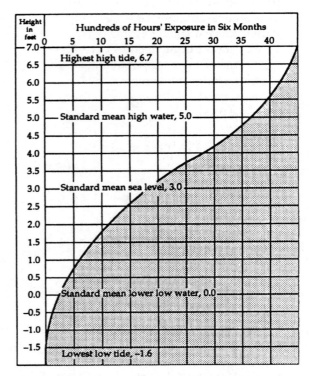

Figure 6.1 (Source: Reprinted from Between Pacific Tides, Fifth Edition, by Edward F. Ricketts, Jack Calvin, and Joel W. Hedgpeth, revised by David W. Phillips, with the permission of the publishers, Stanford University Press. Copyright 1985 by the Board of Trustees of the Leland Stanford Junior University.

tolerate many hours of exposure and others are able to tolerate only a few hours per week or month, the animals and plants sort themselves into subzones within the intertidal zone itself. These distinct subzones are customarily numbered as zones 1–4, land-to-sea (see Figure 16.7 on page 421 in Garrison, or Figure 16.22 on page 363 of Ingmanson & Wallace). The zones have an obviously different appearance even to the person not familiar with shoreline characteristics. Their boundaries are often strikingly sharp because of the astonishing number and variety of life forms they support.

One reason for this phenomenal diversity and success is the relatively large quantity of food found in the intertidal zone. The junction between land and ocean is a natural sink for living and once-living material. Nutrients running off the land from rains serve as rich fertilizers and sources of inorganic nutrients for the inhabitants of the zone as well as for the plankton of the area. The larval forms, which are abundant, and the adult attached organisms frequently use the plankton as their primary food source. The constant crashing of the surf ensures a high concentration of dissolved gases to support a rich plant population. Recent research has indicated that when nutrients are present, wave-stressed plant communities may yield ten times the productivity of a tropical rain forest! Intense wave action apparently facilitates nutrient flow, optimizes the distribution of light on photosynthetic tissue, speeds gas exchange, and deters grazers. These plants are also responsible for the protection of a large larval fauna of oysters or fish or other economically important organisms.

Another reason for the success of organisms here is the number of habitats and niches provided within the whole intertidal zone. The addresses of the animals and plants vary from very hot, intensely saline "splash pools" to cool, dark crevices under rocks. These spaces provide hiding places, attachment sites and jumping-off spots, cracks from which to peer to obtain a surprise lunch, footing from which to launch a surprise attack, or darkness to shield a retreat.

Of course, not all intertidal areas are rocky. Some shores are sandy, some are gravel, some are mud, and some are made of rounded rocky fragments called "cobbles." Some shores even combine all these elements within a very small area.

The muddy-bottomed salt marshes are among the most interesting of the nonrocky shores. They are filled with plants that not only survive but prosper in a salt water marine environment. Occasionally these salt marshes form in estuaries where fresh water and salt water mix. These estuarine marshes are even richer and exhibit greater species diversity than marshes exposed only to pure seawater. The primary productivity in these areas is extraordinarily high, and the standing biomass (mass of living matter) is among the highest per unit of surface area of any inhabited environment.

Estuaries are sometimes called oceanic nurseries because of the large number of juvenile forms found there, especially those of fish. Many pelagic varieties spend their larval lives in the protective confines of an estuary.

A particularly interesting contrast between rocky and sandy intertidal areas and salt marshes is that in the former nearly all the species are marine. A greater mix of terrestrial and marine species exists in the salt marsh, indicating the truly transitional nature of this unusual and productive habitat.

Marine animals that live in the intertidal zone fit into one of two groups. The motile animals, such as crabs, snails, and flatworms, move about to find their food. Sessile animals, such as mussels and barnacles, are attached to the substratum or, in some instances, to other animals. The ocean brings their food to them. No comparable way of life exists for animals on land.

The organisms that live in the intertidal area face severe problems. Wave shock is probably the most awesome. Whole rocky breakwaters have been moved tens of feet by wave action. Rocks have been thrown nearly 300 ft into the air by the force of breaking waves. The motile animals hide, and the sessile animals hang on, often gaining assistance from their rounded shells that fend off forces around their bodies. Sessile animals in rocky intertidal areas may have a special glue with which they can firmly attach to a rock, or a foot that insinuates into small cracks to provide a good hold. In sandy beaches the animals burrow deep into the sand or move seaward out of the impact zone.

Desiccation (drying) is another source of stress. Again, the motile organisms have the advantage because they can move toward the water left in tidal pools by the retreating ocean, but sessile animals and plants must patiently await the water's return. Often, water trapped within the shell keeps gills moist for the needed exchange of gases, or a mucous coat retards evaporative water loss from blades of seaweed. When the weather is too warm or the tide is out for an unusually long time, a sleeplike quiescence settles over the zone to be relieved only by the return of the life-giving sea.

Learning Objectives

After completing the reading assignment and viewing the program, the student will be able to:

☐ Understand the basic physical characteristics of the intertidal zone.

☐ Name four different types of intertidal shores.

☐ List the four intertidal subzones, and briefly describe the difficulties of living in each.

☐ Differentiate between motile and sessile organisms.

☐ Describe some of the biological modifications required for life in the intertidal zone.

Key Terms and Phrases

Intertidal (Littoral) The marine zone between the highest high-tide point on a shoreline and the lowest low-tide point.

Tidal zero The reference plane from which tidal height is measured. Although this reference plane varies in different countries because of arbitrary decisions on what should be called zero, the Pacific Coast now uses a running average of the height of the lower of the two low waters (low tides) that occur through a day's time. So we have as our reference plane the mean lower low water. When you see that a tide will be 0.0 at 1134, that means that the water will reach the reference point at 11:34 a.m.

Zones (Subzones within the intertidal zone) The distinct biological communities, often easily recognized by gross appearance, that make up the intertidal area. Customarily listed from 1 to 4, land-to-sea.

Larva (Plural is *larvae*) A young organism that becomes self-sustaining and independent before it takes on the characteristics of its parents. (From text glossary)

Motile Able to move about.

Sessile Unable to move freely. Sessile organisms are anchored to substratum by glue, cords, and so on. Parts of the organism may be able to move, but the organism as a whole is anchored in place.

Wave shock Physical movement, often violent and of considerable force, caused by the crash of a wave against an organism.

Desiccation Drying.

Estuaries The coastal mouth of a river valley where marine influences predominate. Generally a very productive environment. Water typically brackish and subject to tidal effects.

Primary productivity The amount of organic material synthesized by organisms from inorganic substances, nearly always accomplished by photosynthesis.

Before Viewing

☐ Read "Physical Factors Affecting Marine Life" on pages 334–337, and pages 414–427 in Chapter 16 of Garrison; or "Salinity Fluctuations: Life in Estuaries" and "Interactions Among Organisms: Life Along the Shore" on pages 362–367 of Ingmanson & Wallace.

☐ Note the diversity and complexity of a typical intertidal situation as illustrated on pages 420–421 of Garrison, or pages 364–365 of Ingmanson & Wallace.

☐ Watch Program 6: "The Intertidal Zone."

After Viewing

Reread the text sections, being sure you understand the key terms and phrases, and complete the following exercises:

1. Considering all of the places where living things can exist in the ocean environment, which would you consider to be the most difficult?

2. If the intertidal zone is such a rigorous place to make a living, why are there so many organisms living there?

3. Why are estuarine salt marshes so productive, and why do they contain so many species?

Answer Key for Exercises

1. *Difficulty* in biology is a relative term. It may seem a circular argument, but wherever organisms live, conditions for life at that place are biologically tolerable, food is available, and environmental conditions are not so extreme as to preclude success. The most difficult area, considering the number and violence of the physical factors that must be routinely tolerated, is surely the intertidal zone. And the sand beach is certainly the most rigorous intertidal area. Because of the shifting sand base, the great difficulty of separating food from substratum, the ease with which predators can discover the location of prey, the capillary forces that compact the grains, the exposure to abrasion and wave shock, and tens of other factors, the sand beaches surely are prizewinners in this category. But what is the most difficult beach? Black sand beaches derived from pulverized lava in the Hawaiian area, surely. In addition to all the other difficulties mentioned, we must add the lava's ability to store solar heat until temperatures approach 160°F just below the surface of the sand. Almost nothing can live on these beaches for more than just a few minutes, including human feet!

2. Organisms live in abundance where energy is available. Energy, in the form of accessible food, is readily available in the intertidal region. Where there is food, there is life. Evolutionary adaptation has sorted out the ways that work in this zone from the ways that do not, and the ways that work give the organisms living there access to a rich harvest of nutrients.

3. Lots of reasons. For one thing, rivers can be a source of mineral nutrients and enriched sediment. As the river expands and slows, much of this nutrient material settles to the bottom. For another, the special shape of an estuary often protects the area from wave shock, and usually some sort of protecting sand bar system will form at the seaward boundary. The calm, rich surroundings are amenable to a great number and variety of plants and animals from land and sea. The regular ebb and flow of the tides brings larval recruits from the ocean as well as additional sources of food. All things considered, an estuary can be a very pleasant place for huge numbers of organisms to live.

Optional Activities

1. One of the most startling sights for someone new to marine biology is the first view of a tide pool. If you have never visited this habitat, you have a pleasant surprise and a delightful couple of hours in store. Be sure to take a guidebook of some sort (see Supplemental Reading).

2. Many beginning aquarists stock their tanks with intertidal species because they are very hardy animals. If a seawater aquarium store or display is in your area, you might be able to see some of the intertidal organisms mentioned in your text and in the television segment even though you live many miles inland. Note: Virtually all intertidal habitats in the United States are now marine preserves protected by law. You must not carry anything out of these areas unless you have a valid collector's permit and have a good scientific reason for disturbing the animals or plants. Even uninhabited shells and rocks are protected in many areas. When you turn a rock over to investigate its hidden side, be sure to return it to its previous position. Watch where you step. Be gentle. Leave the organisms of the shorelines for others to enjoy as well.

Self-Test

1. The intertidal zone nearest to shore is traditionally numbered
   a. zone 1
   b. zone 2
   c. zone 3
   d. zone 4
   e. zone A

2. Generally speaking, the intertidal area is
   a. rich in life, but only a few species live there
   b. rich in life, with considerable species diversity
   c. not particularly rich in life
   d. not particularly rich in life, but rich in food
   e. virtually devoid of life because of the difficulty of surviving there

3. Organisms can protect themselves from wave shock by
   a. hiding when a wave comes
   b. gluing themselves to the rocks
   c. sliding into a small crack as the wave crashes around them
   d. running away
   e. all of the above

4. Perhaps the most difficult oceanic habitat is
   a. the abyssal zone
   b. the intertidal zone in general
   c. the intertidal rocky shore
   d. the intertidal beach
   e. the intertidal black sand beach

5. Which of these statements is *not* true of estuaries?
   a. Larvae are often abundant there.
   b. Estuaries are in danger of development and pollution.
   c. Very few plants live in estuaries, but many animals can be found there.
   d. Estuaries are places where salt water and fresh water meet.
   e. Estuaries are scoured by tidal currents moving in and out.

6. Food in the intertidal zone is
   a. not particularly abundant
   b. abundant but inaccessible
   c. abundant and accessible, but no organisms are able to take advantage of it
   d. abundant and accessible, and many animals are able to take advantage of it
   e. none of the above

7. The Tidal Datum for the West Coast of the United States is
   a. mean low water
   b. mean lower low water
   c. mean water
   d. mean high water
   e. mean higher high water

8. The intertidal zone with the least number of species is
   a. zone 1
   b. zone 2
   c. zone 3
   d. zone 4
   e. zone A

9. The greatest species diversity can be found in
   a. zones 1 and 2
   b. zones 1 and 4
   c. zones 3 and 4
   d. zones A and 3
   e. zones A and B

10. Which of these intertidal areas is generally considered to be the most productive and to be inhabited by the largest number of species?
    a. rocky intertidal zones
    b. mud flats exposed to 100 percent seawater
    c. salt marshes associated with estuaries
    d. gravel flats
    e. sand beaches

Supplemental Reading

Hinton, S., *Seashore Life in Southern California.* Berkeley: University of California Press, 1969.

Horn, M. H., and R. N. Gibson, "Intertidal Fishes." *Scientific American*, Jan. 1988, 64–70. (Fishes in the intertidal zone have special adaptations for clinging, and some species can breathe air.)

Koehl, M. A. R., "The Interaction of Moving Water and Sessile Organisms." *Scientific American*, vol. 247, no. 6 (1982), 124.

Lewin, R., "Life Thrives Under Breaking Ocean Waves." *Science*, vol. 235, no. 6 (1987), 1465–1466. (Information on the unexpectedly high productivity of intertidal plant communities.)

McConnaughey, B., and R. Zottoli, *Introduction to Marine Biology*. St. Louis: Mosby, 1983.

Ricketts, E. F., and J. Calvin, *Between Pacific Tides*, Fifth Edition. Stanford, Calif.: Stanford University Press, 1974.

Lesson Seven   Continental Margins

Overview

When you are standing on the beach with a tentative toe testing the cold water, it may seem that you are at land's end, about to step off a continent into an ocean basin. But if you could sail out to sea and peer into the waters below, you would discover that the continent extends seaward many miles as a submerged platform called the continental shelf. The shelf is part of the continental margin, a dynamic transition zone that also includes the continental slope and the continental rise. These boundaries between land and sea have fluctuated widely over geologic time in response to motions of the earth's great crustal plates. The location of the shoreline today is just a small part of an ever-changing cycle.

The continental margins are being explored at present as never before, their lures being the great potential resources such as oil and gas, minerals and fisheries. Although the margins are varied, they can be classified into three categories based on motions of crustal plates.

Active margins are generally narrow, characterized by extensive faulting, frequent earthquakes, active volcanoes, and deep sea trenches known as "subduction zones." The rugged mountainous west coasts of North and South America are active margins, a result of collision or convergence of the leading edge of the westward moving continents with the adjacent ocean basins.

The east coast of North America is a *passive* margin. It has a broad continental shelf as much as 100 mi wide, is seismically quiet (lacking volcanoes and deep trenches), and is gently sloping like the bordering coastal plain. Passive continental margins form on the trailing edge of the westward-moving continent and are related to divergent plate boundaries, such as the

opening of the Atlantic Ocean. In the Arctic Sea, off the coast of Siberia, the continental shelf reaches a width of about 1,280 km (800 mi), which is astonishing indeed, considering the worldwide average is only about 68 km (42 mi).

The third type of continental margin is the *translational* margin, characteristic of the southern California coast. This section of the continental shelf, known as the southern California Continental Borderland, is within a very active seismic zone. Intense faulting has fractured the platform into a series of high islands (such as Catalina), shallow submerged banks, and deep elongated basins. The margin has been influenced by the primarily horizontal motion along the famous San Andreas Fault, which lies about 50 mi inland. The fault is the boundary between the northwest-moving Pacific Plate and the westward-moving North American Plate.

The average depth at the outer edge of continental shelf worldwide is about 130 m (426 ft) but can be as great as 500 m (about 1,640 ft) on glacially eroded shelves. The average slope is about 1 ft per 1,000 ft, a slope angle of about 0.1°, or about 7 min. (There are 60 min in 1 degree of angle.) The smallest slope that the human eye can detect is about 17 min; therefore, except for local irregularities, the continental shelf would appear to be as flat as a billiard table.

The submerged shelves today receive most of the rock debris from the erosion of the continents. Deep soundings indicate thousands of feet of accumulated sediments and sedimentary rock piled on the edge of the continents. The sedimentary rocks are the source of the oil and gas produced off California, the Gulf Coast, and the North Sea.

The surface of the continental shelf was greatly altered during the emergent periods of the last Ice Age. The shelf has been submerged such a short time, geologically speaking, that the relic deposits of the glacial period are still present, affected only slightly by waves and recent sedimentation. Old sand dunes, muddy marshes, peat bogs and rounded pebbles of ancient beaches are frequently dredged from portions of the shelf now under water. Hills of rocky glacial debris mark the terminal moraines of continental glaciers that pushed their way on to the exposed land. U-shaped troughs and valleys gouged out by glaciers etch conspicuous irregularities in the shelves of the high latitudes. Steep V-shaped valleys are evidence of the rivers that flowed over the exposed shelf, lengthening their courses to reach the lowered sea.

As the climate eased and sea level rose, the coral reefs that thrive today on many island and continental shelves grew upward in the warm tropical waters. The near-shore waters are habitats of high biologic productivity. The finest fishing grounds are in shallow shelf waters, and mariculture projects of the future will probably be established in these locations.

Today, the forces that influence the continental shelves are varied, but the three most important are waves, rivers that contribute sediments, and faulting that occurs on every shelf. Tides and tidal currents, the occasional tsunami or seismic sea wave, storm surges, and certain burrowing marine organisms can also alter features of the shelf.

Beyond the edge of the continental shelf is the continental slope, a drop-off that forms the confining rim of the ocean basins. Near the deep-sea trenches that circle the Pacific Basin the slope may plunge from 3 to 6 mi – one of the most dramatic topographic features of our planet. If we could stand at the bottom of the Peru-Chile Trench that parallels the west coast of South America (with a maximum depth of about 25,280 ft) and peer upwards along the west wall of the Andes (rising in places to over 22,000 ft), we would see a scarp soaring over 47,000 ft! It would look like Mt. Whitney (the highest peak in the continental United States) piled on top of Mt. Everest (highest peak in the world) with another 1,220 m (4,000 ft) to go.

Not everywhere is the scenery so dramatic, but nowhere is the continental slope merely a smooth featureless scarp. It is incised by numerous valleys and canyons trending down slope with the canyon walls cut by smaller gullies. Recent interest in the economic potential of continental margins has uncovered new information about the features of the shelf and slope, as well as the canyons. Canyons have been

discovered on all of the continental shelves around the globe – in passive and active margins, cut both in solid rock and in sediments. They have been found in various shapes and sizes – short and long, straight and branched – with some approaching the Grand Canyon in depth and form.

On land most canyons are created by stream erosion. However, other processes must be operating in the sea. The most likely explanation for the formation of submarine canyons is that the upper parts of the canyons were first eroded by streams flowing across the exposed continental shelf during the Ice Ages when the sea level was lowered. During the decline of the Ice Ages when the ice was melting and the sea level rose again, canyon development may have continued through erosion by underwater landslides; slump; or dense currents of sand, mud, and rock moving down slope. This process continues today. The sediments that are eroded from the slope and canyon walls are transported seaward to be deposited at the mouth of the canyon or on the continental rise as great fans of unconsolidated material. Some of the fan deposits show graded bedding, indicating deposition by turbidity or density currents.

At the base of the continental slope are the great heaps of unconsolidated sediment called the continental rise. The rise marks the seaward edge of the continental margin and overlies the junction between the continents and the deep seafloor, particularly in the Atlantic. This sedimentary apron has not been well explored, nor is the underlying structure between the edge of the continental crust and the ocean crust clearly understood. In much of the Pacific Basin the continental rise is essentially missing, and the junction may be a trench, a subduction zone where slabs of oceanic crust are apparently diving under the nearby continent.

The margins of the continents reflect the forces acting both on and within our planet. As landmasses are wrenched apart by movements in the crust and mantle, new continental shelves will form, opening new niches and habitats for marine organisms. This is the time when many new species are developed and the abundance of marine life increases.

But what happens to the margins when continents collide? It is believed that the great mountain ranges such as the Appalachians, the Alps, even the lofty Himalayas, were once continental margins later shoved upward by the collisions of landmasses.

As we study the continental margins of today and seek out the wealth of petroleum hidden in the deep sedimentary layers, one question persists. How will these pull-aparts and collisions affect life in the future?

Learning Objectives

After completing the reading assignment and viewing the program, the student will be able to:

☐ Sketch a diagrammatic cross section showing the continental and oceanic crust, and indicate the continental shelf, slope, and rise.

☐ Describe the features found on continental shelves; list the processes that influence the shelf and relate the appropriate structures to each.

☐ Identify the salient features of the continental slope and rise.

☐ Compare and contrast active, passive, and translational continental margins, using examples from the coasts of the United States.

☐ Describe submarine canyons and relate current theories of origin.

☐ Discuss the effects of the Pleistocene Ice Age and the changing sea levels on the continental margins.

Key Terms and Phrases

Igneous rock Any rock formed from the solidification of molten material. Igneous rocks are the primary rocks of the earth from which all other rocks are derived.

Volcanic or extrusive igneous rock forms from *lava* that cools on the earth's surface, including the seafloor. Basalt is the most common extrusive igneous rock.

Intrusive igneous rock forms from *magma* (or molten material) that cools at depths within the earth's crust. Granite is the most common intrusive igneous rock.

Continental crust The land surface composed primarily of granite with a thin veneer of sedimentary rock. The average thickness of the crust is about 50 km and the average density is 2.7 g/cm^3.

Oceanic crust The ocean basins, which are made of basalt with a thin cover of organic and inorganic sediments. The average thickness is about 7 km and the average density is about 3.0 g/cm^3.

Granite An intrusive igneous rock that is light-colored and coarse-grained with a characteristic black and white (salt and pepper) appearance. Granite is composed primarily of quartz and feldspar and is considered a *sialic* rock because of the abundant *si*licon and *al*uminum. It is the primary constituent of the continental basement rock that may be uplifted by earth movements and exposed in the core of mountainous areas such as California's Sierra Nevada range. Granite is rarely encountered in the ocean basins.

Basalt An igneous rock that may be intrusive or extrusive. It is dark-colored and dense, composed primarily of *si*licon, *ma*gnesium, and iron minerals and, therefore, considered a *simatic* rock. The basement rock of the ocean basins and oceanic islands such as Hawaii is basalt. Basalt is not limited to oceanic crust and is widespread on the continents.

Ice Age or Pleistocene period Period of fluctuating temperatures during the last million years or so, during which time the ice covered the continents four times and retreated with warm interglacial periods between. Study of the Ice Age is important to oceanographers because it was a period when sea temperatures cooled and sea level was lowered 100–200 m. During this time the continental shelves were dry land, as evidenced by fossil land plants and animals recovered by coring shelf sediments.

Continental shelf dams Offshore structures, such as volcanoes, faulted and uplifted basement rock, coral reefs, and salt domes that may trap sediments and hold them against the continents, extending the area of the shelf.

Submarine canyons Steep-walled, narrow canyons cut into the continental slope and shelf. They are generally perpendicular to the continental margin. Submarine canyons are found worldwide, cut into all types of continental and island shelves to depths over 800 m. They appear to be erosional features resulting from underwater landslides, turbidity currents, or slumps of unconsolidated material. They may be fault controlled, although the origin of such faults is still not well understood.

Delta A deposit of river-transported sediment (primarily quartz sand and clays) on a continental shelf. Most large rivers have deltas that are generally triangular in shape like the Greek letter delta (thus, the name). The size and shape of a delta will ultimately depend on the amount of sediment carried by the stream, as well as by the waves and currents at the site of deposition. Strong longshore currents tend to move sediment deposits along the beach, preventing delta formation.

Before Viewing

☐ Read Chapter 4, "Continental Margins and Ocean Basins," in Garrison, or Chapter 6, "Margins of the Continents," in Ingmanson & Wallace. Read also "Characteristics of U.S. Coasts" on pages 311–312 of Garrison, or "Coasts of the United States" on pages 252–261 of Ingmanson & Wallace.

☐ Study Figure 4.9, page 101 of Garrison, or Figure 4.3, page 54 of Ingmanson & Wallace, to distinguish between the granitic rock of the continents and the basaltic rock of the seabed.

☐ Note in Figure 3.13 (Garrison) or Figure 5.13b (Ingmanson & Wallace) the dynamic relationship between seafloor and continents. Note the movement of Earth's mantle.

☐ Read the descriptions of submarine canyons and study the diagrams and photographs to understand the location and origin of these submerged valleys. Study pages 105–106 in Garrison, or pages 87–88 in Ingmanson & Wallace.

☐ Watch Program 7: "Continental Margins."

After Viewing

Complete the following exercises:

1. What are the natural forces at work today modifying and shaping the continental shelves?

2. What role do shelf dams play in the building of continental shelves? Describe the different kinds of shelf dams and tell where they are found.

3. Could you defend the statement "Continental slopes are the greatest cliffs on earth"? In which ocean, Atlantic or Pacific, are the slopes most dramatic, and why?

4. If another Ice Age were to descend on us and the sea level dropped about 125 m, approximately how far would the Port of New York be from the shoreline? See Figure 4.17 on page 105 of Garrison, or Figure 6.9 on page 89 of Ingmanson & Wallace for a clue. If the earth were to enter a warming trend and the glaciers on Antarctica and Greenland were to melt, what might happen to New York, the Florida Everglades, and other coastal areas of the world?

5. If you are using the Ingmanson & Wallace text, study Figure 6.10 on page 90. Note the Columbia River, Astoria Submarine Canyon, and the Astoria Fan. The Columbia is a major river, yet it has not formed a delta on the continental shelf. In fact, the other rivers draining the Coast Ranges and the Klamath Mountains have no deltas either. What factors might account for this?

Optional Activities

1. Examine the contours of the submarine canyons shown in Figure 4.16 of Garrison or Figure 6.7 of Ingmanson & Wallace. If you are using the Ingmanson & Wallace text, about how far from the beach is the head of Scripps Canyon? The letters *nm* stand for nautical miles: 1 nm = 1,852 m (6,076 ft). What is the depth at the bottom of Scripps Canyon? Could this feature have been the result of stream erosion during the Ice Age sea level lowering? Explain. Notice the divers in the canyon. If you were diving in Scripps Canyon, what are some of the features you would see? If you are ever in La Jolla, California, stop in at Scripps Institution of Oceanography and see the fine model of Scripps Canyon in their Aquarium.

2. Study Hudson Canyon in Figure 4.17 of Garrison or Figure 6.9 in Ingmanson & Wallace. How long is this canyon from the shore to the lowest contour shown? What is the depth at the bottom? Any ideas on the origin of this spectacular feature?

Self-Test

1. Which of the following statements does *not* apply to the ocean basins?
   a. The basement rock is granite.
   b. The oceanic crust is thin, about 7 km in thickness.
   c. The basement rock has a relatively high density.
   d. The sediment cover is thin.
   e. The rocks of the seafloor are simatic.

2. Igneous rocks are described as
   a. having solidified from molten material
   b. the primary rocks of the earth
   c. extrusive if the molten material solidifies on the earth's surface
   d. intrusive if the molten material solidifies within the crust
   e. all of the above

3. The continental shelf
   a. is a narrow strip on the East Coast of the United States
   b. is a steeply dipping zone dropping off to the deep-sea floor
   c. is a featureless plain unlike the neighboring continent
   d. is a gently sloping platform with a variable landscape
   e. all of the above

4. Most of the sediments eroded off the continents will be deposited
   a. in the submarine canyons
   b. on the abyssal plains
   c. on the beach
   d. on the continental shelf
   e. in offshore sand bars

5. Which of the following occurred during the Pleistocene period of extensive glaciation?
   a. The sea level was lowered 100 to 125 m and the continental shelves were exposed as dry land.
   b. The sea level rose 150 m.
   c. Many land areas we see today were inundated with water.
   d. The continental shelves were under deep water.
   e. Shells and bones of deep-water organisms were deposited on the continental shelves.

6. The *most* important natural forces influencing the continental shelves today are
   a. longshore currents
   b. turbidity currents and river currents
   c. waves, rivers, and faults
   d. tsunami and storm surges
   e. all of these

7. Which of the following statements accurately describes active continental margins?
   a. They are regions of great geologic stability.
   b. They are characteristic of the margins of the Atlantic Basin.
   c. They are areas of frequent earthquakes and volcanoes, where crustal plates are converging or are in collision.
   d. They are areas where crustal plates are actively moving apart.
   e. They are usually different in topography from the adjoining coast.

8. The transition zone between the shelf and the seafloor is
   a. the littoral zone
   b. the continental slope and rise
   c. the abyssal plain
   d. the mid-ocean ridge
   e. the submarine canyon

9. Continental shelf dams are important because they
   a. prevent waves from attacking a shore
   b. tend to hold the continents together
   c. prevent large predatory fish from coming into shallow water
   d. hold continental sediments on the shelf and build out the shelf
   e. can be used to generate hydroelectric power

10. Submarine canyons
    a. are found worldwide, on all kinds of shelves
    b. are steep-walled and narrow
    c. are cut into solid rock
    d. are cut into the shelf and may extend almost to the beach
    e. all of these

11. The origin of deltas is related to
    a. glacial deposition and the formation of moraines
    b. river deposition of sediments eroded from continents
    c. glacial erosion and the formation of troughs and fjords
    d. volcanic activity in coastal regions
    e. biological activity of corals and certain shelled organisms

12. Which of the following statements apply to thick carbonate deposits on continental shelves?
    a. They appear off the coasts of Texas and Florida.
    b. They are characteristic of warm shallow seas.
    c. They are found around volcanic islands and atolls in the tropics.
    d. They were a common occurrence in the past.
    e. All of the above apply.

Supplemental Reading

Burchfiel, B. C., "The Continental Crust." *Scientific American*, vol. 249, no. 3 (1983), 130–142.

Emery, K. O., "The Continental Shelves." *Scientific American*, vol. 221, no. 3 (1969), 106–122.

Francheteau, J., "The Oceanic Crust." *Scientific American*, vol. 249, no. 3 (1983), 114–129.

McGregor, B. A., "The Submerged Continental Margin." *American Scientist*, vol. 72, no. 3 (1984), 275–281.

Lesson Eight   Beyond Land's End

Overview

It is quiet there and bitter cold. The pressure of four miles of water above is crushing. The blackness is broken only by flashes of living light from animals attracting prey, or perhaps a mate. This is the realm of the deepest sea, the abyss.

For eons, the deep sea was undisturbed by events taking place in the upper waters. The greatest waves churned up by the winds do not stir either the sediments or the waters of the deep ocean. The changing of the seasons passes unnoticed, and even on the hottest days in the tropics, the deep water stays near freezing. The circulation seems sluggish, nothing like the currents and eddies of the surface layers. Even the organisms of the abyss seem unaware of the activity in the warm turbulent waters of the upper zones.

Within the last decade or two, all has changed. The stillness of the abyss has been shattered by swarms of probes and water samplers, by holes drilled into the seafloor, by dredges scraping the bottom, and by sonic vibrations mapping the details of the underwater landscape. The flashing lights are now likely to be underwater cameras. People in capsules dive deeply to visit this alien world, then scurry back with tales of incredible wonders. This new study of "inner space" is an exciting pursuit, equaling in interest our probing of the planets and the cosmos.

Unexpectedly, these studies have revealed a seafloor different from the familiar landscape of the continents. The rugged midocean mountains are the longest chains on earth, rising from the floor of every major ocean and interconnected like the seams on a baseball. These ridges and rises are not made of the folded and eroded sedimentary rocks seen in the Alps, the Himalayas, or the Appalachians, but are piles of igneous black basalt that welled up from the earth's crust and mantle and congealed in the cold water. The central rift valley, characteristic of all midocean ridges, is a result of separation or pulling apart of two crustal plates. The eruption of the basalts from fissures or volcanic cones within the rift has over millions of years created the long ridges and rises and the new seafloor that makes up the ocean basins.

Innumerable single volcanic peaks dot the floor of the basins, particularly in the Pacific, sometimes poking their heads above water to form islands, but usually standing as seamounts, not tall enough to break the sea surface.

The valleys or trenches of the seafloor are also different from their counterparts on land. They are not the products of erosion, as is the Grand Canyon, but elongated troughs that are being pulled or pushed downward deep into the earth's crust. Most trenches circle the Pacific on the seaward side of arc-shaped chains of volcanic islands. The trenches are narrow, generally V-shaped, and are the deepest parts of the ocean basins. The islands bordering the trenches are volcanically active, but do not produce basalt, the rock of the seafloor and the midocean ridges. These more explosive volcanoes emit andesite, a lava of somewhat different chemical composition.

Some of the most fascinating features of the deep seafloor, the hydrothermal vents or hot springs of the midocean rises and ridges, were discovered nearly 20 years ago. In 1977, a team of scientists diving in a research submersible near the Galápagos Islands were the first to visually observe and sample this remarkable environment. Since that time – using underwater cameras, probes, and the deep diving submersible *Alvin*, and working at depths as great as 3 km (1.8 mi) – oceanographers have located other springs and chimneys where waters emerge at temperatures as high as 350°C! The hot waters

41

bring up ores of iron sulfide and other valuable minerals, which are deposited around the edges of the hot springs.

The sulfide-rich water supports what may be the strangest biologic habitat on earth. The astonished scientists who first saw the eerie landscape around the vents found giant clams, huge eight-foot tube worms with blood-red gills, limpets, crabs, strange stalked animals resembling dandelions, and many others. Vast growths of chemosynthetic bacteria apparently provide food for the clams, the worms, and some of the other animals in the habitat. Here, in the absence of light and photosynthesis, in waters warmed by the rising magma of a spreading seafloor, life is sustained by chemical reactions very different from anything on land or in the surface waters. (See Lesson 23.)

Another recent discovery that makes us question our concept of the abyss as a region of calm water is evidence of "storms" of swift flowing bottom currents, three miles below the surface and skirting the western edge of the ocean basin. These disturbances, which have been detected by current meters moored to the Atlantic seafloor, are not always present, nor do they occur everywhere. They do, however, scour the seafloor in some places and transport and deposit huge volumes of sediment in others.

The slow-moving circulation of deep cold waters from the polar seas has been long known and well charted. The onset of the abyssal storms, however, is unpredictable and does not seem to follow a seasonal pattern. The rapid current lasts about a week before it dissipates.

Knowledge of these abyssal storms is of particular interest to engineers who are responsible for cables and structures placed on the seabed. They must also consider the effect of bottom currents on toxic waste that may be dumped on the ocean floor. Should the wastes be contained in one area or spread about? Obviously, further research is needed; it may show us more is going on in the deeps than we imagine at present.

Although the rough lava surface of the ocean floor has not been gentled by erosion, it has been smoothed and modified by multicolored carpets of sediments, unconsolidated materials derived from various sources. The most widespread deposits in the deeper part of the ocean are the fine-grained muds and clays, usually brick red, chocolate brown, and occasionally dark olive green in color. These clays

resting on the basalt basement are derived from the land and are the end products of slow, continuous weathering of continental rocks. The particles are transported to the sea primarily by rivers and slowly settle out of the water to the seafloor. Near the continents and the mouths of large rivers, sediments may measure tens of kilometers in thickness and may accumulate at the rate of several meters per year. Far from land, however, where only the finest clay particles will be carried out to sea, the rate of accumulation may be only 1 or 2 mm per thousand years.

In shallower waters and on the tops of the ridges and rises, a creamy white deposit rests like a blanket of snow. Microscopic examination of this light-colored "ooze" reveals countless tiny shells of planktonic foraminifera, diatoms, or radiolarians that have rained down from the surface to be preserved on the seafloor. To be classified as an ooze, a deposit must consist of at least 30 percent organic remains. Ancient oozes, dried and compacted, are found in many places on the land of today, evidence of previous periods of flooding. The calcareous deposits of foram shells make up the chalks of the White Cliffs of Dover and the sticks of chalk found in every school classroom. The siliceous oozes, usually made of shells of diatoms, small single-celled plants, are prized as diatomaceous earth, the main constituent of swimming pool filters and countless other products.

Deep-sea photographs and dredging have revealed another widespread deposit, especially on the floor of the Pacific: the enigmatic manganese nodules. Spread out like a field of sooty black potatoes, the nodules apparently are not brought in from the land like the red clays, nor have they fallen through the water like the oozes. Their exact origin is still a matter of controversy, although many oceanographers feel they formed in place on the seafloor and may be related to volcanic activity. In addition to manganese, these nodules contain appreciable percentages of iron, copper, cobalt, and nickel, all metals that will someday be in short supply on land. But they rest in about 4,570 m (15,000 ft) of water and are difficult to extract. And who owns them, out there in the middle of the ocean?

The seafloor sediments provide powerful clues to many aspects of earth history. By examining the layers in deep cores, we can see the changing forms of marine organisms through the ages. We can trace ancient ocean currents, plot long-range climatic changes, and locate islands (that may literally have blown themselves into oblivion) from the remains of their volcanic debris. In addition, ocean sediments contain valuable deposits of oil and gas, manganese, phosphate, and other commercial minerals.

But the sedimentary record also brings up questions that have wrenched at the very foundations of scientific theory. We know that the earth is about 4.6 billion years old and that the waters are almost as old. We note that the oceans are the great catchment basins of the planet and should contain the sedimentary record of all events through geologic time. Yet in all of the coring and dredging done so far, rocks recovered are no older than 200 million years. The ocean floor seems to be a relatively youthful feature, and the oldest rocks encountered are not in the sea but on the continents. Where is the original seafloor with its cover of the very first sediments? Have the waters been sloshing around from one basin to the next, one ocean closing as another opens?

The abyss is a deep and vast area and does not give up its secrets easily. But oceanographers are persistent, and the light of knowledge is beginning to illuminate even the darkest sea.

Learning Objectives

After completing the reading assignment and viewing the program, the student will be able to:

☐ Locate on a world map the major topographic features of the ocean basins, including the major oceans, the larger marginal seas, the midocean ridges and rises, the trenches, and selected islands and island chains.

☐ Describe the characteristics of the major geologic features of the ocean basins.

☐ Compare the origin of seafloor landforms with their counterparts on the continents in terms of processes operating.

☐ Appreciate Darwin's interpretation of the origin of coral reefs and atolls.

☐ Classify the various kinds of seafloor sediments, know their special properties, and compare their modes of origin.

☐ Plot the distribution of the major seafloor sediments on a world chart and give reasons for this occurrence.

☐ Identify the methods used by oceanographers to learn about the ocean basins and the seafloor sediments.

Key Terms and Phrases

Abyss The deepest portions of the sea, generally below 3,700 m.

Oceanic ridges Seafloor mountains, present in every ocean basin, made of basaltic volcanic rock erupted through fissures and vents in the central rift valley of the range.

The highest mountains in the Atlantic average more than 3,000 m above the seafloor, with peaks exceeding 4,000 m. The flanks, crests, and central valley of the midocean ridges are major topographic features of the earth's surface. The crests rise as much as 1,000 m on either side of the valley floor.

Oceanic valleys (1) Rift valley between the crests of the midocean ridges and rises; a zone of earthquake activity, active volcanism, bounded by faults, high heat flow.

(2) Deep-sea trenches. Elongated troughs or depressions in the seafloor parallel to continental margins such as the Peru-Chile Trench, or to island arcs, such as seaward of the Aleutians, Philippines, Japan, Java. Sites of deep earthquakes. Deepest parts of the sea.

Pillow lavas A widespread feature of the seafloor formed by the extrusion of molten lava from fissures or vents into rounded tubes or "pillows" generally resembling toothpaste. Pillow lavas are seen in the rift valley of the Mid-Atlantic Ridge, near the volcanic islands and other vents.

Abyssal plain Vast, flat, featureless areas of the seafloor usually covered with thick unconsolidated sediments. Abyssal plains are more widespread in the Atlantic and Indian Oceans than in the Pacific.

Lithogenous or terrigenous sediment Seafloor deposits composed of particles from the weathering and erosion of the land; transported to the oceans by streams, wind, or glaciers. These sediments consist of fine muds or clays, fine quartz grains, windblown dust, and glacial debris in high latitudes. Volcanic ash may be locally abundant. They are classified by particle size and shape, density of the mineral grains, color, and chemical composition. Red and brown clays are the most widespread terrigenous sediment of the deep sea.

Biogenous sediment Organic deposits or oozes made of the microscopic shells of various single-celled planktonic organisms. Calcareous oozes are made of the calcium carbonate ($CaCO_3$) shells of foraminifera, pteropods, and coccolithophores. They are characteristic of temperate waters less than 4,000 m deep.

Siliceous oozes are composed of shells of single-celled plants called diatoms and shells of radiolaria, both planktonic. Diatom oozes are characteristic of the high latitude seafloor surrounding the Antarctic and across the North Pacific. Radiolarian oozes are found in the Equatorial Pacific Basin.

Authigenic deposits Deposits that apparently form on the seafloor by precipitation out of seawater. Manganese nodules are an interesting example of an authigenic deposit. They are widespread on the Pacific floor in deep water and consist of golf ball to fist-sized lumps – potato-shaped, black, brown, or gray – sitting on top of the sediment and showing an internal layered structure. Potentially valuable as a source of iron, manganese, nickel, copper, and cobalt. Phosphate nodules are an example of a shallow-water authigenic deposit and are used in parts of the world as sources of calcium phosphate, a fertilizer.

Turbidity currents Swift flows of dense mud, silt, sand, and sometimes fossils of shallow-water organisms onto the deep seafloor. Turbidity currents also include underwater landslides from the continental shelf, which travel down continental slopes and through submarine canyons. They are triggered by earthquakes, storms, tides, internal waves, and floods and are an important method of bottom transport of sediments to the deep sea. They form turbidite deposits, which exhibit graded bedding from coarse particles on the bottom to fine-grained materials at the top.

Bottom currents Strong bottom currents that have flowed consistently for thousands of years, primarily on the western side of the ocean basins, and have produced great drifts and ridges of sediment, mud waves, current ripples, furrows, and other bed forms of the deep sea.

Before Viewing

☐ Reread Chapter 2, pages 38–39, "The Sampling Problem," in Garrison. Continue with a review of Garrison's Chapter 4, "Continental Margins and Ocean Basins," then read Chapter 5, "Sediments." Finish with the brief description of hydrothermal vents on pages 430–432.

☐ If you are reading Ingmanson & Wallace, reread Chapter 3, pages 30–51, then read Chapter 4, "Ocean Basins and Sediments." Finish with the material on pages 26–28, and pages 361–362, on hydrothermal vents.

☐ Using the world map at the end of either text, locate the principal topographic features of the ocean basins: the Mid-Atlantic Ridge, the East Pacific Rise, the Mid-Indian Ridge, the abyssal plains, the deep-sea trenches that ring the Pacific, the Puerto Rico Trench, the Hawaiian Islands, the Aleutian Trench, and the Aleutian Islands (off Alaska). Find the Galápagos Islands near the discovery site of the hydrothermal vents.

☐ Using Figure 5.13 on page 134 of Garrison, or Figure 4.13 on page 66 of Ingmanson & Wallace, match the distribution of sedimentary deposits to the topographic features of the seafloor that you just observed in the world map at the end of either text.

☐ Watch Program 8: "Beyond Land's End."

After Viewing

Reread the text sections. Using the world map, try to identify areas discussed in the television program. Reread the key terms, especially those referring to sediments. Locate each type on Figure 5.13 on page 134 of Garrison, or on Figure 4.13 on page 66 of Ingmanson & Wallace. Complete the following exercises:

1. Imagine you could drain away the water and take a walk on the deep-sea floor. Describe an abyssal plain, such as the Hatteras Abyssal Plain in the Atlantic.

2. Compare and contrast the features of the Mid-Atlantic Ridge with a continental mountain range such as the Rockies or the Appalachians. What are some features seen

in the central rift valley of the ridges and rises? Where can oceanographers see the rifted zone on dry land?

3. Describe the Pacific Ring of Fire.

4. Take an imaginary dive in a submersible into the Mariana Trench. Give a running report of the landscape as you cruise along the bottom. Is anybody living there?

5. Take an imaginary dive in the *Alvin* to a hydrothermal spring. What will you see there not usually seen on the open seafloor?

6. Glacial marine deposits consist of pebbles, boulders, and gravels that have fallen to the seafloor from melting icebergs. Where do you think these deposits have been encountered? Are any glacial marine sediments forming today? Where?

7. If you were to put a calcareous ooze under a microscope, what kinds of organisms would you see? How would they differ from those in a siliceous ooze near Antarctica? See Figure 5.10 on page 130 of Garrison or Figure 4.10a on page 63 of Ingmanson & Wallace. Where were these shells most likely found? In addition to oozes, what kind of sediments are widespread on the seafloor?

8. How do manganese nodules form? What do they look like? Why are so many nations interested in their recovery? What are some of the problems in mining manganese nodules?

Self-Test

1. The midocean mountains
   a. are found in every ocean basin
   b. are composed of igneous rock
   c. differ in origin and structure from continental mountains
   d. have a rift valley between the central crests
   e. all of these

2. The greatest thickness of sediments on the seafloor is located
   a. in the rift valleys of the midocean mountains
   b. on the high peaks of the midocean mountains
   c. on the seamounts and guyots
   d. on the abyssal plains
   e. in the littoral zone

3. The most widespread terrigenous deposits on the seafloor are
   a. the calcareous oozes
   b. glacial marine deposits
   c. red and brown clays
   d. siliceous oozes
   e. volcanic ash

4. Volcanic eruptions under the sea form
   a. granite boulders
   b. oozes
   c. turbidites
   d. pillow lavas
   e. diatoms

5. Oceanographers classify the mysterious manganese nodules as _____ deposits.
   a. calcareous ooze
   b. lithogenous
   c. hydrogenous or authigenic
   d. siliceous
   e. organic

6. The volcanoes in the island-arc trench systems generally erupt
   a. basalt lavas
   b. calcium carbonate rock
   c. andesites
   d. continental granites
   e. all of these

7. Which of the following descriptive phrases do *not* apply to turbidity currents?
   a. sluggish slow-moving streams
   b. bottom transport of sediments
   c. a dense mixture of rocks, silts, and sands
   d. a type of underwater landslide
   e. may be set off by earthquakes, storms, and other disturbances

8. The oozes on the seafloor mostly consist of
   a. boulders and cobbles from glaciers oozing off the land
   b. bones and teeth of bottom-living fish
   c. fine muds washed down the continental slope to the seafloor
   d. microscopic shells of single-celled surface-living organisms
   e. treated sewage wastes from urban areas

9. Which of the following metals is *not* usually found in manganese nodules?
   a. iron
   b. uranium
   c. nickel
   d. cobalt
   e. copper

10. The landscape within the central rift valleys of the midocean mountains
    a. is marked by active volcanoes
    b. is covered by pillow lavas
    c. has steep faulted cliffs
    d. is bounded by high mountain crests on either side
    e. all of these

11. Great drifts of sediments, mud waves, and ripple marks indicate
    a. strong tidal action
    b. large wind waves
    c. eruptions of lavas on the seafloor
    d. strong bottom currents
    e. active faulting

Supplemental Reading

Edmond, J. M., and K. Von Damm, "Hot Springs on the Ocean Floor." *Scientific American,* vol. 248, no. 4 (1983), 78–93.

Haymon, R. M., and K. C. Macdonald, "The Geology of Deep-Sea Hot Springs." *American Scientist*, vol. 73, no. 5 (1985), 441–449.

Hekinian, R., "Undersea Volcanoes." *Scientific American*, vol. 251, no. 1 (1984), 46–55.

Oceans, "The Abyss." Vol. 12, no. 6 (1979). (The following articles are included in this issue: "In Utter Darkness" by J. E. McCosker; "All About Ooze" by N. Rosa; "Benthic Currents" by I. Ashkenazy; "Voyage to the Bottom of the Sea" by Don Walsh; "Angus: Eyes in the Deep" by David Clark.)

Lesson Nine   Plate Tectonics

Overview

About 400 years ago, when the margins of the Atlantic were becoming better known, a few curious people pondered the almost uncanny similarity of the opposing coasts of Africa and South America. Although the early charts were far from accurate, the margins of the continents could be fitted together like pieces of a giant jigsaw puzzle to form one supercontinent. The idea that the continents might have once been joined simmered for a long time, but as more information slowly accumulated, the possibility seemed to become less and less likely. The continents and other ocean basins are solid rock, and the first studies of earthquake waves indicated that the crust of the earth and the underlying mantle are rigid also. The beautiful fit of the shores of the Atlantic was relegated to a coincidence, maybe even a cosmic joke, for there was no feasible way to move such large masses through such a solid rocky crust.

But the notion of a supercontinent still surfaced from time to time as new discoveries were made. In the early 1900s, Alfred Wegener, a meteorologist, not only noted the match of the continental margins but also called attention to the similarity of fossil plants and animals found on either side of the Atlantic. Also, when the continents were fitted together, the trend of certain ancient mountain ranges continued neatly across the boundaries. Wegener proposed the theory of drifting continents to explain the evidence he had gleaned from many fields of science. He named his one great continent Pangaea and suggested this land began to break into individual blocks about 150 million years ago. But Wegener could not demonstrate a satisfactory method of pushing his continents through the solid crust, so his theories were accepted by only a handful of geologists, derided by most.

Despite this, a revolution was on its way. Great vertical displacement of rock could be seen everywhere, in high mountain ranges, in uplifted plateaus, and in downwarped basins. These movements suggested that over long periods of time even solid rock could slowly flow under conditions of extreme heat and pressure.

In the last few decades, as discoveries and data pour in and scientific papers pour out, few scientists resist the theories of seafloor spreading and plate tectonics. The picture of a firm earth with fixed patterns of land and sea has been replaced with the notion of a not-so-"firma terra," having a crust made of seven large plates, several smaller plates, and possibly many miniplates that slowly move about, slipping on a hot, nonrigid layer of partially molten rock in the upper mantle. There are indications that continents have repeatedly collided and joined, only to break apart and separate in different patterns, while new ocean basins formed in the rifts. The Atlantic Ocean is our most obvious example of a young ocean still pulling apart, with new seafloor forming from the hot lavas pouring out of the central rift valley of the Mid-Atlantic Ridge.

Today most oceanographers are enthusiastic "drifters," for the overall theory is well documented, although many details are yet to be explained. This revolution in thought came about quickly once the data were in, because, like all good theories, it is a unifying concept that offers a better, simpler, more logical explanation for observed phenomena. We now understand why earthquakes occur in linear belts. These belts are the plate boundaries where plates are separating along the midocean ridges, converging near the deep-sea trenches, or grinding past one another as in California along the San Andreas Fault. Californians know, for example, that when the plates rattle, it is the plates moving.

The deep-sea trenches were another puzzle, for they should have been filled in with sediment washing in from the adjacent continents and island arcs. We believe they are zones of convergence where the oceanic crust is thrust downward into the mantle. The deep-focus earthquakes, the recorded low heat flow, and the low gravity measurements in the trenches fit well with our model of a diving, cold, oceanic slab. We can now explain why the oldest rocks on earth, over 3.8 billion years old, are found on the continents, while nothing older than about 200 million years has been found in the ocean basin. All the seafloor rocks have been recycled down the trenches.

Since the earliest sailors used magnetite (lodestone) to navigate, it has been known that the earth has a magnetic field with magnetic north and south poles that will influence iron particles and compass needles. The origin of the earth's magnetic field is probably related to the earth's hot and fluid outer core and the relative motions of the solid inner core and the mantle. The magnetic north pole is not at the same location as the geographic or rotational North Pole but is inclined about 11.5° away. The compass needle, therefore, does not point to true north but to magnetic north. The variation in degrees between true and magnetic north will depend on the latitude and longitude of the observer. All navigational charts include the magnetic variation of each locality, so compass deviation can be corrected during the voyage. In addition, the position of the magnetic north pole changes slightly each year, a factor that must also be noted by the long-distance navigator.

Studies of layered lava rocks and sedimentary deposits reveal another interesting fact about the earth's magnetism. The direction of the magnetic field apparently reverses from time to time; many such changes in polarity have occurred during geologic time. When molten lavas have erupted and started to cool, magnetite crystals start to form, aligning themselves with the earth's magnetic field like tiny pointers toward the magnetic poles. When the lava solidifies, the oriented magnetic particles are a permanent record of the magnetic field at the time of eruption. From this record seen in rocks of different ages, we see evidence of change in the location of the magnetic poles, variations in the intensity of the earth's magnetic field (which

is now weakening), and of reversals of the polarity of the magnetic field.

When studies were made of the Mid-Atlantic Ridge just south of Iceland, a peculiar pattern of alternate linear bands of strong and weak magnetism appeared in the seafloor rocks. The width of the magnetic bands was similar on either side of the central rift valley, with strong and weak anomalies in matched pairs equal distance from the crest of the ridge. It was reasoned that the strong magnetic readings matched the present orientation of the earth's magnetic field; the weak bands were from rocks that have reversed polarity. As the oceanic crust was created in the central rift valley, the seafloor separated and the older magnetized rocks moved away from each other. The new lavas that welled up to fill the gap took on the magnetic orientation of the earth at that moment, at times the reversed field. This process has been continuing for at least 200 million years in the Atlantic, with polar reversals occurring at irregular intervals as the seafloor spreads apart. The magnetic readings, like a tape, record both the magnetic events and the growth of the seafloor. Magnetic striping has been discovered in every ocean basin, and, combined with methods of age-dating seafloor rocks, has become the "clinching" argument in favor of plate tectonics.

These are exciting times for the earth scientist as the study of the mobile crust continues in the ocean basins and, more recently, on the continents. Plate tectonics is a powerful tool, helping us unravel the complex history of our planet.

Learning Objectives

After completing the reading assignment and viewing the program, the student will be able to:

☐ Discuss the development of theories from continental drift and seafloor spreading to modern plate tectonics.

☐ Understand the significance of earthquake belts and magnetic striping of the seafloor as evidence for plate motion.

☐ Relate other lines of evidence that support the theory of plate tectonics.

☐ Outline on a world chart the major plates and indicate their present motion.

- ☐ Compare and contrast the three types of plate boundaries, their location and special features.

- ☐ Describe the origin of the Hawaiian Islands in terms of the Hot Spot theory.

- ☐ Realize the implications of living on a plate boundary.

- ☐ List some of the unsolved problems regarding plate tectonics and suggest lines of research.

Key Terms and Phrases

Tectonics The study of the history, structure, and motion of the earth's crust and upper mantle.

Continental drift The theory proposed by Wegener that suggests the continents are in motion, based on evidence of matching continental margins in the Atlantic and other scientific data.

Pangaea Wegener's supercontinent prior to about 200 million years ago and the opening of the Atlantic basin.

Seafloor spreading The mechanism by which oceanic crust is generated at spreading centers or zones of divergence, such as the rift valleys in the midocean ridges.

Crustal plates The six or seven large and several small blocks of the earth's crust and upper mantle, each moving as a separate rigid unit, bounded by regions of active earthquakes.

Asthenosphere A zone in the upper mantle, extending from about 100 km to 200 km below the earth's surface over which the crustal plates slide. The rocks in the asthenosphere are believed to be plastic or partially molten, based on the decrease in the velocity of seismic waves at this depth.

Plate boundaries Many of the more dramatic geologic phenomena, such as mountain building, opening of new ocean basins, volcanic eruptions, active faulting, and major earthquakes, take place at the boundaries between moving crustal plates. The plates may pull apart from each other, collide, or grind past each other, in each case producing different effects and features.

Divergent plate boundaries A region where crustal plates are spreading apart, such as in the central rift valleys of the midocean mountains. The rift is steep-walled, about 1 km deep, and ranges in width from about 12 to 48 km. The crests of the ridges are riddled with faults, and shallow depth earthquakes are frequent. Cracks and fissures within the rift zone emit lavas, forming cones, pillow lavas, sheet flows, and other volcanic structures. The youngest rocks of the seafloor are found at divergent boundaries and, in places, may be "today old." This zone is where new seafloor is being formed from the molten magma rising from the earth's mantle.

Convergent plate boundaries Regions where crustal plates are colliding or converging. Usually when two plates converge, one is forced under, resulting in the formation of deep-sea trenches, deep-focus earthquakes, and volcanic activity.

Ocean-continent convergence occurs where dense basalt ocean crust collides with and subsides under less dense granite of the continents. The Peru-Chile deep-sea trench that parallels the west coast of South America is the "subduction zone" where the seafloor is plunging under the westward moving continent. Earthquake waves trace the path of the diving plate as it descends into the hot asthenosphere.

Ocean-ocean collision of two dense plates will result in the subduction of one plate into a deep-sea trench. As the cold diving plate heats up, the rock becomes partially molten, erupting lavas that will eventually build a volcanic island. The island arc and trench system of the western Pacific Ocean and the Aleutian Islands are the result of ocean-ocean convergence.

Continent-continent collision is between masses of low-density rock; neither will subduct or subside. Both plates will become compressed, folded, faulted, and uplifted. The mighty Himalayas are still growing today as India presses north into Asia. Many folded mountain ranges, such as the Alps, are the results of convergence of former continental margins.

Lateral or transform plate boundaries A zone where crustal plates are slipping past each other, such as along the famous San Andreas Fault in California. The Pacific Plate is moving northwest, carrying Los Angeles (west of the fault) toward San Francisco (east of the fault). Portions of the San Andreas are slipping quietly, but parts seem to be caught; there, the strain seems to be building. Both Los Angeles and San Francisco are located near these sections of the fault.

Subduction zone An inclined slab of oceanic crust descending into the mantle, forming the deep-sea trenches. The Benioff Zone, a region of deep earthquakes, marks the zone between the diving oceanic plate and the overriding plate.

Spreading rates Range from 1 to 2 cm/yr along the Mid-Atlantic Ridge to about 10 to 12 cm/yr along the East Pacific Rise. Rates as high as 20 cm/yr have been estimated for parts of the South Atlantic when it first began to rift apart.

Transform faults and fracture zones Result in horizontal displacement of the central rift valleys of the midocean mountains. One of the largest, the Romanche Fracture Zone, offsets the axis of the Mid-Atlantic Ridge by almost 1,000 km. The vertical relief within the Romanche Zone, from the crests of the ridge to the troughs, is over 6 km, about four times greater than the Grand Canyon. Great vertical displacement of the seafloor and oceanic crust is indicated by the presence of peridotite dredged from the fracture zone. Peridotite is considered to be the chief constituent of the earth's mantle rock.

Before Viewing

☐ Read Chapter 3, "Earth Structure and Plate Tectonics" in Garrison; or Chapter 5, "Plate Tectonics" in Ingmanson & Wallace. Note especially Table 3.1, page 79, "Characteristics of Plate Boundaries" in Garrison, or Table 5.2, page 79, "Summary of Plate Tectonic Interactions" in Ingmanson & Wallace. Pay careful attention to the illustrations in the chapter – tectonic processes are dynamic and are best described visually.

☐ In the Glossary of the text, look up the following words or expressions:

continental drift   transform faults
lithosphere   transform boundary
magnetic anomaly   translation margin
seafloor spreading

☐ Watch Program 9: "Plate Tectonics."

After Viewing

Using the world map at the end of either text, locate the features discussed in the text and the program. Reread the text sections, being sure you understand the key terms and phrases, and complete the following exercises:

1. Describe the features on the seafloor and on the continents that strongly support the modern theory of plate tectonics.

2. Explain the causes of magnetic striping of the seafloor.

3. Identify the six or seven major crustal plates and indicate the direction of motion of each.

4. Locate the boundaries of the major plates on the world map at the back of the text and determine which are convergent, divergent, or translational.

5. Describe what you would see if you were in the submersible *Alvin*, descending into the rift valley in the Mid-Atlantic Ridge. What does the landscape look like? Locate the Romanche Fracture Zone on the world map at the back of the text. What features would you see there? What is the significance of the rift valley in plate tectonic theory? What is the significance of the deep-sea trenches in the theory?

6. It is said that North America will move westward about your body length in your lifetime. If North America drifts at about 2 cm/yr, is this a reasonable estimate? (2.5 cm = about 1 in.)

Optional Activities

1. Examine Figure 3.17 on page 74 of Garrison, or Figure 5.2a–c on page 70 of Ingmanson & Wallace. Try to project the future configuration of the continents and ocean basins about 50 million years from now.

2. What will happen to the Atlantic Basin? What will become of the Red Sea and the Gulf of California? Trace the travels of India. Do you think they are over? Follow the movements of Australia. Where will it move next?

Self-Test

1. Land-based evidence for continental drift can be seen in
   a. the distribution of Glossopteris fauna and flora
   b. evidence of ancient glaciation
   c. the lineation of mountain ranges
   d. the correlation of rocks in now widely separated continents
   e. all of these

2. The youngest seafloor rocks are
   a. found nearest the continental slopes
   b. found nearest the rift valleys of the mid-ocean ridges
   c. found under the deep-sea trenches
   d. evenly distributed over the ocean basins
   e. underlying the continental shelves

3. Crustal plate boundaries may be
   a. a deep-sea trench
   b. a rift valley
   c. a belt of earthquakes
   d. a lateral or translational fault
   e. all of these

4. The magnetic striping of the seafloor is considered evidence of seafloor spreading and
   a. subduction down the rift valleys
   b. spreading centers in the trenches
   c. changes in the earth's axis of rotation
   d. reversals in the polarity of the earth's magnetic field
   e. periodic collapses of the earth's gravitational field

5. The subduction zones are important to the theory of plate tectonics as regions
   a. of new seafloor formation
   b. of plate divergence or pull-apart
   c. of plate convergence, where old seafloor is being "consumed"
   d. of great stability in the center of the plates
   e. all of these

6. The oldest rocks on earth have been found
   a. in the deepest parts of the sea
   b. on the continents
   c. in the midocean ridges and rises
   d. in the trenches
   e. on volcanic islands, particularly Surtsey

7. The boundary between the North American plate and the Pacific plate can be seen in the
   a. trench off southern California
   b. Mariana Trench
   c. Rocky Mountains
   d. San Andreas fault zone
   e. East Pacific Rise

8. According to modern plate tectonics theory
   a. only the oceanic plates are moving around
   b. only the continental plates are moving
   c. the earth's crust and upper mantle are slipping on the asthenosphere
   d. the earth's crust and mantle are slipping on the molten outer core
   e. only the ocean waters are moving from one basin to another

9. Of the following, which is *not* considered part of a young rifting sea?
   a. the Red Sea
   b. Gulf of California
   c. Atlantic Ocean
   d. Iceland
   e. the Mediterranean Sea

10. New earth crust is being generated today
    a. in the deep-sea trenches
    b. in submarine canyons
    c. in the rift valleys of the midocean ridges and rises
    d. in the centers of the continents
    e. under the continental rise

11. The force driving the crustal plates is believed to be
    a. earth magnetism
    b. gravity
    c. the pull of the sun and the moon
    d. convection cells in the upper mantle
    e. deep-water currents of the ocean basins

Supplemental Reading

Bass, I. G., "Ophiolites." *Scientific American*, vol. 247, no. 2 (1982), 122–131.

Bonatti, E., "The Rifting of Continents." *Scientific American*, vol. 256, no. 3 (1987), 96–103.

Bonatti, E., and K. Crane, "Oceanic Fracture Zones." *Scientific American*, vol. 250, no. 5 (1984), 40–51.

Fryer, P., "Mud Volcanoes of the Marianas." *Scientific American*, Feb. 1992, 46–52. (Mantle rock, transformed into mud by water distilled from the subducting Pacific plate, oozes up along faults near the Mariana Trench to form mountains on the seafloor.)

Hoffman, K. A., "Ancient Magnetic Reversals: Clues to the Geodynamo." *Scientific American*, vol. 258, no. 5 (1988), 76–83.

Hsu, K. J., "When the Black Sea Was Drained." *Scientific American*, vol. 238, no. 5 (1978), 53–63.

Jones, D. L., A. Cox, P. Coney, and M. Beck, "The Growth of Western North America." *Scientific American*, vol. 247, no. 5 (1982), 70–84.

Jordan, T. H., and J. B. Minster, "Measuring Crustal Deformation in the American West." *Scientific American*, vol. 259, no. 2 (1988), 48–58.

Macdonald, K. C., and B. P. Luyendyk, "The Crest of the East Pacific Rise." *Scientific American*, vol. 244, no. 5 (1981), 100–116.

Molnar, Peter, "The Structure of Mountain Ranges." *Scientific American*, vol. 255, no. 1 (1986), 70–79.

Murphy, J. B., and R. D. Nance, "Mountain Belts and the Supercontinent Cycle." *Scientific American*, April 1992, 84–91. (The authors believe supercontinents have formed repeatedly in a tectonic cycle that lasts about 500 million years.)

Nance, R. D., T. R. Worsley, and J. B. Moody, "The Supercontinent Cycle." *Scientific American*, vol. 259, no. 1 (1988), 72–79.

Oceanus, "Deep Sea Hot Springs and Cold Seeps." Vol. 27, no. 3 (1984). (An excellent issue concerned with the geology, physiology, and biochemistry of the deep-sea vents and hot springs.)

Powell, C. S., "Peering Inward." *Scientific American*, June 1991, 100–111. (Excellent review article outlining controversial new views of the Earth's interior.)

Sclater, J., and C. Tapscott, "The History of the Atlantic." *Scientific American*, vol. 240, no. 6 (1979), 156–174.

Lesson Ten Islands

Overview

When the pressures of the world lie heavy upon us, who has not thought of moving to an island, an isolated speck of land, replete with tropical fruits and flowers, surrounded by azure seas, and tinted by scarlet sunsets? Gauguin's magnificent paintings of Tahiti intensify this dream of escape. To the oceanographer, islands have their own particular fascination, rising as they do thousands of feet from the deep-sea floor, or projecting as rocky promontories on the flat platform of a shallow continental shelf.

In spite of variations in size, shape, geography, age, and origin, islands have certain common attributes. They are all relatively small land masses, surrounded by water, and more or less isolated from the continents. The endemic plants and animals that inhabit the islands may be found nowhere else on earth. Remember Darwin and the wonderful animals of the Galápagos? Most islands are found in the Pacific, although they occur in every ocean basin and marginal sea.

Some islands are temporary features, appearing on nautical charts and disappearing before the next batch of explorers reaches the spot. Those islands that are more or less permanent and well-documented are so numerous and of such great variety that classification is a monumental task. A broad organizational scheme has been devised based on origin and rock type.

The largest and most familiar islands are called continental islands because they are made up of continental rocks, such as granites and sedimentary rocks. These islands are usually close to land on a continental shelf and were probably once part of the mainland. Although they have been separated for varying lengths of time by changes in sea level and tectonic move-

ments, they usually maintain plants and animals similar to those on the adjacent continents. The number of species may be greatly reduced, however. The British Isles have about half the species found on nearby mainland France. Ireland has about two-thirds of those seen on neighboring Britain. These islands separated from the mainland in the recent geologic past, and some of the organisms were unable to migrate across the water. The best known examples of continental islands are the British Isles, New Zealand, New Guinea, the Seychelles in the Indian Ocean (far from land but part of an ancient granite basement), and Madagascar, also in the Indian Ocean.

Another kind of island found near continental margins is the barrier island found on sandy coasts near flat continental shelves. Barrier islands are widespread and generally develop in two different ways. Sand can accumulate offshore on broad shelves where the waves are small. The sand bar that is formed by wave action will build until it is above sea level. Plants will then colonize the new land, partially stabilizing it against wave erosion.

Barrier islands can also be built by a longshore current moving sand across the mouths of lagoons and bays. Islands formed by this process can be seen offshore from New Jersey to Florida. These barriers protect the coast and coastal facilities against erosion by waves. But they are often vulnerable and short-lived because they are formed of unconsolidated sand. Sometimes, during high tides and major storms, waves will wash over the low-lying islands, destroying the structures, even cities, built on these scenic but unstable lands.

Off southern California lies a string of small islands called the Northern Channel Islands. These islands are similar in structure and rock type to the adjacent mountains on the mainland. Because they are separated from the mainland by a 20-mi-wide, deep basin, controversy continues over whether they were ever connected to the continent. The puzzle is enhanced by the discovery of dwarf mammoth bones in Ice Age deposits on the islands. Did these strange, extinct mammals walk or swim to their island home?

The Florida Keys, which extend southerly from the tip of Florida, have a very different origin from that of other continental islands. Florida, including the continental shelf, is part of an ancient carbonate bank. The Keys, which are small, mangrove-covered islands, are just the peaks of a fossil reef built on the gently sloping shelf. The reef consists of coral and algal debris, shells of fossil mollusks and other warm-water organisms, cemented together by reprecipitated limestone (calcium carbonate). The islands stand only slightly above sea level, susceptible to damage from storms, hurricanes, and high waves. The marine life and colorful birds attract many visitors to the Keys and the nearby Everglades.

The second major category of islands is the true oceanic islands, usually volcanic in origin and made of basalt, the rock that forms the floor of the ocean basins. The land area is often small, involving just the tops of volcanoes that were built up from the seafloor by millions of years of lava flow. On the island of Hawaii, the volcanic summits of Mauna Loa and Mauna Kea stand almost 30,000 ft above the seafloor, rivaling lofty Mt. Everest in grandeur. Island volcanoes that are dormant or extinct undergo long periods of erosion by streams and waves to the point that they are eventually planed down to sea level.

Oceanic islands may be in linear chains (the Hawaiian Archipelago, for example), single peaks, or groups of peaks. Examples include the Hawaiian Islands, Samoa, and Tahiti in the Pacific, and Tristan da Cunha and Iceland in the Atlantic.

Another category of islands takes in the island arc-trench systems that ring the Pacific Basin, particularly along the western margins. On the convex seaward side running parallel to the island arcs are the deep-sea trenches that separate the true basaltic ocean basins from the true granitic continental crust. Among the islands in this category are the Aleutians, Kurils, Japan, the Philippines, and of course the Marianas, bordered by the deepest trench on earth, the Mariana Trench, plunging over 11,000 m (36,000 ft) below sea level.

We have learned that the trenches are an important part of plate tectonic theory, and the origin of the bordering arcs of islands is related to the convergence of great crustal plates. In fact, the island arc-trench systems are complex zones where the seafloor is being thrust or pulled downward along a plane that dips under the islands. The frequent earthquakes that occur in these areas tend to be shallow beneath the neighboring trenches and deep under the islands, consistent with the concept of a subduction zone. The pattern is variable, however, and at the eastern end of the Aleutian Islands two parallel island arcs have formed, one volcanic and the other sedimentary. The Andes Mountains on the west coast of South America are bordered by the deep Peru-Chile Trench, but the andesite islands that were apparently offshore in the past are now part of the mountain chain.

Some of the most active volcanoes on earth are found on these arcuate chains and are characterized by explosive eruptions of lava, ash, and pumice, and hot clouds of gases and water vapor. The lavas of the island arcs form a rock called andesite, intermediate in mineral composition between the oceanic basalts and continental granites.

In the ocean basins, literally thousands of peaks on the seafloor do not project above sea level. They are called seamounts. Seamounts have been charted by means of sonar and are usually cone-shaped with fairly steep slopes. According to numerous dredgings and samplings, they are volcanic in origin and are composed by basalt. Some are in chains that seem to be related to fracture zones in the seafloor along which molten material can seep upward and form undersea volcanoes. Others may have formed on a moving seafloor over "hot spots" that provide the molten lava. Still others may be youthful and not fully grown or may have become dormant very early in their existence. Many are believed to have stood higher than they are today but became extinct and subsided with the settling of the seafloor.

Cones that seem to be truncated are called tablemounts or guyots. Guyots have about the same slopes as the seamounts, but, instead of a peaked cone on top, they have a flat plateau at depths generally between 1,000 and 2,000 m (3,300 and 6,600 ft). There are probably a few hundred guyots as compared to thousands of seamounts. Some oceanographers believe that they were once islands that were beveled off by wave action and later subsided. This theory is supported by the recovery of shallow-water reef materials and shells from guyot platforms now deeply submerged.

Islands are part of the great cycles that shift continents, open new ocean basins, and consume older seafloor. They are rafted with the moving plates from their original volcanic sources, sometimes to be carried down a subducting trench. Islands are ephemeral features of the earth, for none are very old in geologic terms. Volcanoes emerging from the sea are among the most dramatic spectacles on earth. Their demise can be even more spectacular, as was the case in the violent explosion that destroyed Krakatoa. No matter what their history, the lure of islands is unmistakable.

Learning Objectives

After completing the reading assignment and viewing the program, the student will be able to:

☐ Compare and contrast the features and origin of continental islands, oceanic islands, and island arc-trench systems.

☐ Discuss the origin of seamounts and guyots.

☐ Describe the birth of Surtsey, near Iceland.

☐ Understand the problems related to building on barrier islands.

☐ Understand the concepts of island biogeography and the methods of distribution of plants and animals to distant islands.

Key Terms and Phrases

Continental islands Usually large islands on a continental shelf or near a continent, composed of continental granites or sedimentary rocks. They may have been connected to the mainland at one time.

Oceanic islands Islands of volcanic basalt rock that were never part of the mainland. Land areas of these islands may be relatively small because they are just the peaks of underwater volcanoes. They are most common in the Pacific.

Island arcs Elongate arcuate chains of volcanic islands primarily in the north and western Pacific Basin. A zone of active andesite volcanoes and deep earthquakes, they are usually bordered by a deep-sea trench on the seaward side of the arc.

Seamounts Cone-shaped hills on the seafloor, volcanic in origin, that do not rise to the sea surface.

Guyots Flat-topped or truncated cone-shaped volcanic mountains, the tops of which are between 1,000 and 2,000 m (3,300 and 6,600 ft) below sea level.

Surtsey A volcanic island that first appeared off the south coast of Iceland in 1963. The birth and growth of this island were carefully monitored by oceanographers because it is located on a spreading center in the Mid-Atlantic Ridge. Surtsey is very important in understanding events occurring in the rift of an active divergent plate margin.

Syrtlingur A small volcanic island formed near Surtsey in 1965, but completely eroded away by waves in less than one year. This is a good example of an ephemeral island.

Island biogeography The study of animal and plant migrations to isolated islands; also a study of the evolution of endemic or native plants and animals and the effects of introduced species on the resident flora and fauna.

Endemic Plants or animals restricted or peculiar to a locality or an island and the result of long isolation in a particular habitat. Many exotic organisms are examples of island endemism, such as Darwin's finches, the giant Komodo Dragon, the honeycreeper birds of Hawaii, coconut crabs, and possibly the dwarf mammoths of the California Channel Islands.

Introduced species Organisms brought to an island, either accidentally or intentionally, that must compete with the endemic flora or fauna for food and living space in a restricted and sometimes impoverished habitat.

Means of populating distant islands Methods of dispersal include the ability that certain seeds, spores, or rafts of plant debris or seaweeds have to float in seawater. Small animals, such as lizards, amphibians, and occasionally small mammals may be carried great distances on these rafts. The coconut is an example of a seed well-adapted to traveling great distances in water.

Insects, seeds, and spores can also be carried to distant islands by air currents or prevailing winds.

Birds carry seeds and spores in their feathers or digestive tract. Occasionally they will be blown off their regular migration routes and will land on isolated islands, adding to the flora, and sometimes staying to breed and become part of the island bird population.

Before Viewing

☐ In Garrison, reread pages 57–58 describing the birth of a volcanic island, then read pages 306–308 concerning coral islands in the tropics. Reread the information on barrier islands on pages 303–306.

☐ In Ingmanson & Wallace, reread pages 55–56 on the birth of Surtsey. Locate Iceland on Figure 4.5 on page 42. Reread pages 58–61 on coral reefs, island arcs, seamounts, and island chains. Reread pages 246–249 on barrier islands. Read pages 256–258 on Florida and the origin of the Keys.

☐ Review island arcs in Lessons 8 and 9 of this study guide.

☐ Watch Program 10: "Islands."

After Viewing

Review the text sections, being sure you understand the key terms and phrases, and complete the following exercises:

1. Compare and contrast continental islands, true oceanic islands, and island arcs. Discuss location, general size, rock type, and modes of origin. How would you classify Australia and Greenland: islands or continents or in between?

2. What is the origin of seamounts? How do guyots differ from seamounts in shape and possibly in geologic or oceanographic history?

3. Why is Surtsey of particular interest to oceanographers? What happened to little Syrtlingur?

4. Under what circumstances will barrier islands and lagoons form?

5. By what methods are distant islands populated with plants and animals? Why is the introduction of some species devastating to the endemic organisms? Why was the mongoose introduced into Hawaii, and what were the effects?

6. Why is the Santa Cruz Island fox an unusual animal well worthy of study? What methods are being used to study the island fox? How do you think it reached Santa Cruz originally?

Optional Activity

Two interesting ideas have been proposed by island biogeographers. The first states that the smaller the island, the fewer the species of plants and animals it can support. The second states that the farther the island is from the mainland, the fewer the species of organisms that will be found there. If you had to defend these ideas, how would you do it?

Self-Test

1. Which of the following are you *most likely* to find on a true oceanic island?
   a. rocks made of granite
   b. thick sequences of sedimentary rocks
   c. very large endemic mammals
   d. very small land areas
   e. all of these

2. Most of the true oceanic islands can be described by which statement(s)?
   a. They are volcanic in origin.
   b. They were once part of a mainland or continent.
   c. They are located on continental shelves.
   d. They have plants and animals very similar to those found on the continents.
   e. All of the above are true.

3. The island arcs are characterized as
   a. elongated chains of islands
   b. volcanic in origin
   c. bordered by a deep-sea trench on the seaward side
   d. usually erupting andesite lavas
   e. all of the above

4. Long Island, New York, originated as
   a. a true oceanic island
   b. part of an island arc-trench system
   c. part of a terminal moraine deposited by an Ice Age glacier
   d. a resistant granite outcrop
   e. none of these

5. Which of the following statements is (are) true of barrier islands and lagoons?
   a. They are an important coastal feature from New Jersey to Florida.
   b. They are characteristic of gently sloping continental shelves.
   c. They require an abundant source of sand.
   d. They are created by strong longshore currents.
   e. All of the above

6. Surtsey has been of particular interest to oceanographers because it
   a. is a typical tropical island populated by warm-water organisms
   b. formed very quietly and scientists could camp on the shore and observe safely
   c. was the first granite island ever formed
   d. formed within the rift zone of the Mid-Atlantic Ridge and could be observed without descending into the sea in a submersible
   e. was the first island ever to appear and disappear within a few months

7. Coconut palms are very common on isolated islands. This is probably due to
   a. extensive planting by early explorers
   b. air dispersal by large birds
   c. the ability of coconuts to float long distances in seawater
   d. dispersal by coconut crabs, as this is their main source of food
   e. none of the above

8. The little Santa Cruz Island fox is an interesting animal to study because
   a. larger mammals rarely reach islands
   b. foxes do not occur on the nearby continent
   c. it is the only endemic mammal on any island
   d. it evolved from a dwarf mammoth into a fox
   e. it is positive proof that Santa Cruz was once attached to the mainland

9. Seamounts may be described as
   a. rare phenomena of the ocean basins
   b. subsea cones made of volcanic basalts
   c. andesite islands bordering deep-sea trenches
   d. islands populated by plants and animals similar to those on nearby continents
   e. temporary islands, quickly removed by wave attack

10. The Florida Keys are
    a. part of an island arc system
    b. made of coral and algal debris from a reef system
    c. granitic rock extending from the Florida peninsula
    d. remains of ancient andesite volcanoes
    e. none of the above

Supplemental Reading

Burton, R., "Instant Islands." *Sea Frontiers*, vol. 20, no. 3 (1974), 130–138. (This is an interesting account of the eruption on the island of Heimaey near Iceland.)

Case, T. J., and M. L. Cody, "Testing Theories of Island Biogeography." *American Scientist*, vol. 75, no. 4 (1987), 402–411.

May, R. M., "Evolution of Ecological Systems." *Scientific American*, vol. 239, no. 3 (1978), 161–175. (Very interesting article on island biogeographers.)

Shinn, Eugene A., "The Geology of the Florida Keys." *Oceanus*, vol. 31, no. 1 (1988), 46–53.

Lesson Eleven Marine Meteorology

Overview

Oceanic storms! The words suggest immense power and force, dramatic sights and sounds, and considerable danger at sea and on land. Even with today's sophisticated weather satellites, elaborate weather prediction networks, and computer models of atmospheric physical characteristics, we are just as much at the mercy of the elements of weather as were our forefathers in their caves.

Weather is the condition of the atmosphere at a particular time and place. The atmosphere of the earth rarely stays calm for long. We are used to short-term changes as a matter of course, but few people consider the long-term changes that have occurred in the atmosphere of this planet. Our earth's gaseous envelope has changed dramatically through time. Today the earth has a primarily nitrogen (78.08 percent) and oxygen (20.95 percent) atmospheric mix, but in the early eons of life on earth the recipe for air was much different, probably consisting of methane, carbon dioxide, and ammonia – a toxic combination of gases for today's animals and plants. However, as green planktonic plants evolved, so did the atmosphere. Through the last three billion years, the composition of the atmosphere has changed dramatically. Slowly at first, then reaching a peak about 600 million years ago, photosynthesis by these simple plants added vast quantities of free oxygen to earth's envelope of air. Their activity altered the capacity of the atmosphere to support life, and animals were able to evolve. Animals are unable to manufacture the oxygen they require, and until adequate amounts of oxygen were placed in the air by the action of plants, animal life was not able to develop on land.

The forces that shape weather and climate are exquisitely interlocked. Foremost among these is the power source for all weather, the sun – a star at a distance of about 93,000,000 mi from us that provides power for virtually all the weather of the earth. Only about one part in 2,200,000,000 of the sun's radiant energy is intercepted by Earth, and just two-thirds of this light actually reaches the surface, but that amount averages 7×10^6 calories per square meter of surface per day, or a truly staggering 23 trillion horsepower! Much incoming light is stopped by the atmosphere, reflecting back into space from clouds; scattered from ice, droplets, or other particles; or absorbed by water vapor and CO_2. Some light is absorbed or reflected from land, but because most of the earth is covered by water, most of the light strikes the ocean. Whether it penetrates depends greatly on the angle of the strike, sea state (turbulence), the presence of ice cover or foam, and other factors.

About 30 percent of this incident light simply bounces off the earth and returns to space. Of the remaining 70 percent, about 50 percent is absorbed by the earth's land and water surface, and 20 percent by the atmosphere. The 50 percent of solar energy striking land and sea is converted into heat, then transferred into the atmosphere by conduction, radiation, and evaporation. The atmosphere, like the land and ocean, eventually radiates this heat into space. Thus *the total incoming heat* (plus that from earthly sources) equals *the total outgoing heat*, so the earth is in thermal equilibrium, growing neither warmer nor colder. The earth's "heat budget" is in balance.

The earth receives this light and heat energy unevenly because the earth is tilted at 23.5° from the plane of its orbit around the sun (see Figure 11.1 on the next page). In our summer the Northern Hemisphere of the earth receives most of the light and heat because the earth is tilted toward the sun. In our winter the Northern Hemisphere of the earth is tilted away from the

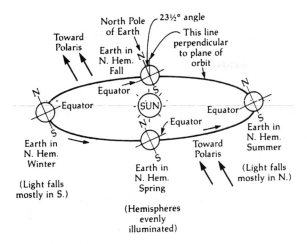

Figure 11.1

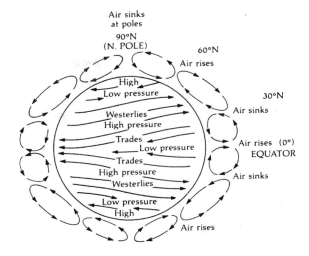

Figure 11.2 Simplified schematic representation of the general circulation of the atmosphere. (Source: From Lutgens & Tarbugck, The Atmosphere: An Introduction to Meteorology, Third Edition, Copyright 1986, page 187. Adapted by permission of Prentice-Hall, Inc., Englewood Cliffs, New Jersey.)

sun, and the Southern Hemisphere receives the greatest proportion of the light and heat. This orbital tilt is the fundamental reason for seasonal changes. Anyone who has noticed that the sun appears to be higher in the sky at noon in the summer than it is at noon in the winter has observed a result of this set of orbital dynamics.

No matter what the season, it is obvious that the equatorial regions of the earth receive more light and heat than the polar regions and are warmer than the poles throughout the year. This difference can be attributed not only to the angle of the sun above the horizon (high overhead in the tropics, low and near the horizon at the poles), but also to light's low angle of approach. Incoming radiation must pass through a greater thickness of atmosphere near the poles, an effect resembling almost constant sunset. The low sun angle and the reflectivity of ice and snow in the polar regions conspire to reflect light and heat back into space.

The air over warm regions tends to become warm and rise, and air over the cold polar areas tends to become cold and fall toward the surface. This warming/rising and cooling/falling cycle moves huge quantities of air across the surface of the earth but does not do so in a single continuous loop from pole to equator. The actual pattern of circulation is suggested in Figure 11.2. The presence of three wind bands in each hemisphere results from the uneven cooling and heating of the air itself, the Coriolis effect, the interaction of air masses, and many other factors. (The Coriolis effect is discussed in detail in

Lesson 12 on oceanic currents and on pages 191–194 of Garrison, or pages 133–135 and Appendix VI of Ingmanson & Wallace.) The bands of winds move around the earth in giant spirals, and surface winds, such as the trade winds or prevailing westerlies, result where the spiral touches the surface. Surface winds blow water along the ocean's surface to cause ocean currents. At the core of these spirals, jet streams can form with winds in excess of 80 m/sec (180 mi/hr), and winds within these streams can move completely around the world in about five days.

Major storms result when masses of air collide or when a single air mass becomes unstable. Air masses are widespread bodies of air that are homogeneous and that take on their major characteristics from the land/sea area beneath them. Two air masses generally do not mix if they collide. The lighter air mass will, instead, be lifted over the more dense one, and because the lifting may cool the upper air mass enough to cause precipitation, rain or snow may follow. Such interactions between air masses occur at a front, and the resulting storm is known as a frontal storm. Frontal collisions are often violent, and waterspouts, thunderstorms, lightning displays, hail, and many other violent weather phenomena can occur.

The most dramatic and powerful weather systems of the earth are not frontal storms but tropical cyclones. These storms are known in the Atlantic as hurricanes, in the Pacific as typhoons, and in the Australian area as Willi-Willis. These huge systems directly influence an ocean area over 400 mi in diameter, have winds in excess of 200 mi/hr, and in some cases form clouds to 50,000 ft above sea level. The sea waves and cloud formation from these storms may indirectly affect entire oceanic and continental areas. Catastrophic rains (up to 15 in./hr) and wind pressure destroy life and property. As they travel over the sea they push ahead a high series of waves known as the storm surge. This storm surge series can create waves up to 40 ft and was responsible in part for the death of some 300,000 people in Bangladesh in the Bay of Bengal in 1973.

The source of power for tropical cyclones is the energy released by the conversion of water vapor into liquid water. The vapor was initially formed by sunlight striking and heating the ocean surface. In a large tropical cyclone the amount of energy released by water condensing is approximately equal to 20 billion tons of flaming gasoline per day, or the explosion of 400 20-megaton hydrogen bombs per day. Such a daily output of energy would be sufficient to power the entire electrical need of the United States for 6 months. And this represents only a very small percentage of the amount of solar energy reaching the tropical oceans in one day!

Will we ever be able to control tropical cyclones? At the moment the prospect does not look likely. As with all major weather systems on the earth, we must be content to be interested and powerless observers.

Another indication of the powerlessness of humanity relates to the heat budget of the earth. An event on April 2, 1982, gave atmospheric scientists a good indication of how sensitively the heat budget of the earth responds to an injection of atmospheric particulate material. The volcano El Chichon (located in Central America) erupted, spewing a huge cloud of dust and sulfuric acid into the stratosphere. Less than one month later, by April 25, the cloud had circled the entire earth in normal atmospheric circulation and was beginning to spread north and south. Stratospheric temperatures subsequently dropped as much as 4°C. Surface insolation rates fell. During this time the great 1982–84 El Niño event began, but meteorologists are unsure of any cause-effect relationship between the two events. The even larger eruption of Mt. Pinatubo

in the Philippines in 1991 is expected to reduce surface temperatures worldwide by as much as 1°C (1.8°F) through the next five years. The magnitude of these changes gives us some idea of the fine balance of the earth's heat budget.

Carbon dioxide also plays an important role in the heat budget. This gas is the product of the combustion of hydrocarbon fuels and is currently building up in the earth's atmosphere. Humanity's unprecedented appetite for energy (oil, coal, and natural gas) is the primary cause for the resulting rise in global temperatures. The overabundance of carbon dioxide may accelerate the "greenhouse effect" in which light converted to heat is trapped near the earth's surface. This phenomenon is analogous to what occurs when light passes through the windows of a car or through a greenhouse pane. The light strikes seats (or plants), is converted into heat, and then cannot escape. On a global average, the year 1990 was the warmest recorded in the past 110 years. Ash and dust from the eruption of Mt. Pinatubo in the summer of 1991 kept 1991 from breaking the record; the first six months of 1991 were the warmest ever recorded. Think what would happen if the earth began inexorably to warm.

Obviously, we can control some things but not others. In this case, to ensure a healthful planet, we should attend to things we can control – for example, our consumption of energy and the processes we use to produce energy – at least until we determine the consequences of our actions.

Learning Objectives

After completing the reading assignment and viewing the program, the student will be able to:

☐ Trace the fundamental change in the composition of the earth's atmosphere since its formation.

☐ Suggest some of the factors influencing weather.

☐ Know the cause of the seasons.

☐ Describe the basic circulation of air in wind bands on the earth.

☐ Differentiate between frontal storms and tropical cyclones.

☐ Note the results of the Coriolis effect on atmospheric circulation.

☐ Understand what is meant by "heat budget."

☐ Suggest some of the effects of natural and human-made pollution on the earth's weather and climate.

Key Terms and Phrases

Weather The condition of the atmosphere prevailing at a specific time and place.

Climate Long-term trends in weather at a location.

Phytoplankton Small, usually unicellular, drifting oceanic plants. (These are discussed in detail in Lesson 15.)

Coriolis effect An apparent force that arises when the speed and direction of a moving object is measured with reference to the surface of the rotating earth. The force is proportional to the speed of movement of the object and acts to cause the moving object to veer off course to the right of its expected direction in the Northern Hemisphere and to the left of its expected direction in the Southern Hemisphere. The magnitude of the force increases from zero at the equator to a maximum at the North Pole.

Latent heat of evaporation The amount of heat required to evaporate a given quantity of water. It is called latent (hidden) because while the amount of heat in the water changes, the temperature of the water does not.

Front The transition zone between two air masses of different temperature, humidity, and density.

Cyclone A low pressure system in which wind blows counterclockwise in the Northern Hemisphere and clockwise in the Southern Hemisphere.

Tropical cyclone A closed, low-pressure center originating over tropical oceans. Tropical cyclones are classified according to their peak wind speeds. Tropical storms, the most highly developed of these cyclonic systems, possess winds exceeding 32.6 m/sec (73 mi/hr). Tropical storms are called hurricanes in the Atlantic, typhoons in the Pacific, and cyclones in the Indian Ocean. These should not be confused with tornadoes or waterspouts, which are small spinning funnels of air usually associated with frontal storms.

Heat budget The sum of the input and output of light and heat energy at the surface of the earth. The input must equal the output over the long term; if it didn't, the earth's oceans would boil or freeze. Distribution of heat from areas of the planet receiving the most heat (equatorial regions) to those areas receiving the least heat (polar regions) is accomplished mostly by the movement of surface ocean currents and by atmospheric circulation.

Frontal storm Precipitation and wind caused by the meeting of two air masses. Generally, one air mass will slide over or under the other, and the resulting expansion of air will cause cooling and, consequently, rain or snow. Frontal storms must not be confused with tropical cyclones, which are storms occurring within one air mass.

Insolation Sunlight falling on the earth.

Before Viewing

☐ Read pages 151–154, "Global Thermostatic Effects," in Chapter 6, and Chapter 8, "Atmospheric Circulation and Weather," pages 185–207 in Garrison, or Chapter 9, pages 124–149, in Ingmanson & Wallace.

☐ Watch Program 11: "Marine Meteorology."

After Viewing

Reread the text section, being sure you understand the key terms and phrases, and complete the following exercises:

1. The earth is actually closer to the sun during the Northern Hemisphere's winter, yet it is colder in this hemisphere during the winter than it is during the summer. How can this be so?

2. The Coriolis effect causes moving objects in the Northern Hemisphere to veer off course to the right (clockwise). Why is it that cyclonic storms in the Northern Hemisphere rotate counterclockwise?

3. Which way does weather tend to move across the United States? Why should this be so?

4. How is heat moved from equator to poles?

5. What effect is pollution having on weather and climate?

Answer Key for Exercises

1. The earth may be marginally closer to the sun due to the characteristics of the orbital ellipse, but the Northern Hemisphere is tilted away from the sun during these months. So, the Southern Hemisphere gets the majority of the heat energy, and the Northern Hemisphere gets a smaller amount. The seasons are caused not by relative distance from the sun, but by orbital tilt.

2. To understand this one you will need to refer to Figure 11.3. Notice that a molecule of air at position 1 begins moving toward the low pressure point in the middle (marked *L*). As it does so, it is influenced by the Coriolis effect and veers off course to the right. The same thing happens to air starting from the position marked 2. The air veers off course to the right as it nears the low pressure area. So does the air starting from position 3. Notice in the case of 3 that the air still moves to the right. You may think the arrow is drifting toward the left side of the page, but that is because you are not observing the wind from the direction of its movement. It may be useful to think of the wind as moving away from your face (as if it had come out of your mouth) and then veering to the right. With these three wind packets moving toward the zone of low pressure and then veering off course to the right, you can see that the air in the immediate vicinity of the low pressure zone will begin to spin counterclockwise because of these tangential forces. So, cyclones in the Northern Hemisphere turn the "wrong" direction for the right reasons!

3. Look again at Figure 11.2. The United States is situated in the zone of winds known as the prevailing westerlies. Weather moves from west to east over the United States.

4. Immense quantities of heat are moved from tropics to polar areas by the action of ocean currents. As you will see in Lesson 12, these

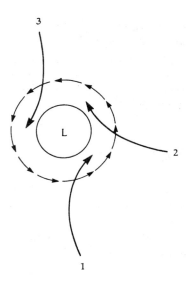

Figure   11.3

currents are, in a sense, huge rivers of water flowing in the oceans. The water is heated by sunlight in the tropics and then moves poleward to warm the coastal areas at high latitudes. Another large-scale mover of heat is the atmosphere. The great circuits of air shown in Figure 11.2 pull on ocean water, causing the ocean currents. Giant tropical storms move air masses from central latitudes to the temperate zones of both hemispheres. Virtually all large-scale heat transfer on this planet is accomplished through the actions of wind and ocean currents.

5. No one is yet certain, though the first reports look ominous indeed. Some researchers say the increased quantities of carbon dioxide, atmospheric dust, and photochemical smogs are causing a significant increase in the retention of heat. The planet's heat budget is being altered – same amount of light in, but less heat out – creating a "greenhouse effect." As the earth heats, the polar caps will melt. Should the great Antarctic ice cap melt, some calculations show that sea levels could rise by as much as 186 ft! Other minority opinions say that the earth's increasing pollution levels will result in less light reaching the surface, and the consequent tilting of the heat budget would be in the opposite direction. This would mean a colder earth. Either way, delicate agricultural balances would be affected; increased rainfall in some areas and droughts in

others would disrupt life on the planet Earth. (We urge you to read the *Scientific American* article by Roger Revelle listed in this lesson's Supplemental Reading section for fascinating reading on this potentially critical situation.)

Optional Activities

1. Most newspapers publish a daily weather map along with predictions for weather for the next 24 hours. Get to know the symbols used in these maps and watch the weather patterns develop for your area from day to day.

2. Weather phenomena are often of great power and beauty. When you see a storm, a rainbow, a waterspout, or any other weather of interest, try to discover its cause and future movements. Often just keeping your eyes open will tell you much about weather and its characteristics.

Self-Test

1. Weather may be defined as
    a. long-term temperature and rainfall trends
    b. long-term humidity and precipitation trends
    c. long-term changes in the composition of the atmosphere
    d. short-term changes in the composition of the atmosphere
    e. short-term, localized characteristics of the atmosphere

2. Seasons are caused by
    a. the varying distance from sun to earth
    b. the wobbling of the earth on its axis
    c. the tilt of the earth's orbital plane
    d. the tilt of the earth's axis with relation to the orbital plane
    e. all of these

3. Air on the earth tends to
    a. rise at the equator and at 30°N latitude
    b. fall at the equator and at 30°N latitude
    c. fall at the equator and rise at 30°N latitude
    d. rise at the equator and fall at 30°N latitude
    e. rise at the equator and at the pole

4. Jet streams can be described by which of the following statements?
    a. They form within wind bands.
    b. They move air at very fast speeds.
    c. They can transport large amounts of air over great distances in little time.
    d. They influence weather in the United States.
    e. All of the above are true.

5. In the Northern Hemisphere there are _____ major wind bands.
    a. two
    b. three
    c. four
    d. five
    e. six

6. Areas of vertical air movement (very little surface wind motion) are found on the earth at latitudes
    a. 15° and 30°
    b. 30° and 60°
    c. 5° and 85°
    d. 0° and 90°
    e. 45° and 55°

7. When air masses come together
    a. very little weather activity results
    b. a tropical cyclone forms
    c. one air mass may ride up and over the other and form a front
    d. one mass may ride up and over the other and form a storm surge
    e. none of the above

8. The power for tropical cyclonic storms comes from
    a. static electricity
    b. the condensation of dry, warm air
    c. the Coriolis effect
    d. the condensation of moist, warm air
    e. the U.S. Department of Energy

9. Which of these does *not* cause loss of life and property in a tropical cyclone?
   a. low atmospheric pressure
   b. wind
   c. rain
   d. storm surge
   e. flooding

10. Tropical cyclones are the same as
    a. hurricanes
    b. typhoons
    c. Willi-Willis
    d. cyclones (as relates to common use in the Indian Ocean area)
    e. all of these

11. As air rises it tends to _____ and occasionally condenses to form clouds.
    a. compress and cool
    b. expand and cool
    c. expand and heat
    d. compress and heat
    e. impossible to predict without more information

12. Which of these is *not* a component of the earth's heat budget?
    a. the amount of sunlight striking the earth (the insolation rate)
    b. the amount of infrared light leaving the earth
    c. sunlight reflected from the earth's oceans, cloud cover, or dust particles in the atmosphere
    d. variations in the "solar constant," the power output of the sun
    e. heat lost from the oceans by evaporation

13. The earth receives _____ of the sun's radiant energy.
    a. about one part in five
    b. nearly half
    c. one part in a billion
    d. about 4 percent
    e. about one part in two billion

14. Which of these atmospheric phenomena is the smallest?
    a. tropical cyclone
    b. waterspout or tornado
    c. frontal storm
    d. jet stream
    e. front

Supplemental Reading

Battan, L. J., *Fundamentals of Meteorology.* Englewood Cliffs, N. J.: Prentice-Hall, 1979.

Ingersoll, A. P., "The Atmosphere." *Scientific American,* vol. 249, no. 3 (1983), 162.

Kerr, R. A., "The Weather in the Wake of El Niño." *Science 240,* 13 May 1988, 883.

McDonald, J. E., "The Coriolis Effect." *Scientific American* (1952).

MacIntyre, F., "The Top Millimeter of the Ocean." *Scientific American,* vol. 230, no. 5 (1974), 62–77.

Oceanus, vol. 29, no. 4 (1986/87). (The entire issue is dedicated to the topic of changing climate and the oceans.)

Rampino, M., and S. Self, "The Atmospheric Effects of El Chichon." *Scientific American,* vol. 250, no. 1 (1984), 48.

Revelle, R., "Carbon Dioxide and World Climate." *Scientific American,* vol. 247, no. 2 (1982), 35. (This is an important article for anyone wishing to understand the effect of humanity on the earth's atmosphere.)

Stewart, R. W., "The Atmosphere and the Ocean." *Scientific American,* vol. 221, no. 3 (1969), 76–105.

Lesson Twelve   Ocean Currents

Overview

During the years he was deputy postmaster general of the American colonies, Benjamin Franklin noticed that certain sailing captains could be depended upon to complete a round trip between North America and England in less time than it took their colleagues. What was particularly surprising was that the speed of the trip often bore little relation to the characteristics of the ship itself; it seemed to depend more on who her captain was! At first Franklin had no luck in discovering the secret of the successful captains. Naturally, these men wished to protect their competitive advantage. But later, Franklin was able to convince these officers and shipowners, one of whom was his cousin, that commerce in general would benefit from this information. They shared their secret.

A "river" flowed in the Atlantic, the captains told Franklin. By altering a ship's course to the north on the way to England, choosing what was actually a longer route, the trip could be made more quickly. The captains watched the temperature and color of the water and stayed within the river on their way east, and were careful to stay south and out of the river on their way back to the west.

Franklin himself had made a number of trips between Europe and North America, and had noticed these differences in water characteristics and trip times. He had even suspected the presence of some sort of river flowing within the ocean itself. After all, the Greek word *ocean* is derived from their word for river. The information from the captains confirmed his suspicions, and in 1777 he published a chart of the Gulf Stream based on these observations.

The Gulf Stream is a fine example of a geostrophic current. These huge water transport systems occur in every ocean. They are respon-

sible for most of the mixing that moves heat and water and organisms from the tropics to the poles and back again. Their influence on the earth's weather is significant, as is their role in fertilizing the upper layers of the ocean with upwellings of nutrient-rich water. Currents are of great importance in oceanography.

Currents are powered by the winds. As we have seen in Lesson 11, winds flow over the earth in definite patterns. These global wind patterns, along with the configuration of continents at this moment in geologic time, have resulted in the formation of a set of geostrophic currents around the circumference of each major ocean. These currents in an ocean are collectively called a gyre.

Gyres in the Northern Hemisphere turn clockwise and those in the Southern Hemisphere turn counterclockwise. This is due to the action of what is called the Coriolis effect, a deflecting force first explained by the nineteenth-century French mathematician for whom it is named. Simply, the Coriolis effect is the tendency for moving objects in the Northern Hemisphere to veer to the right, away from a straight course. Water moving in currents, therefore, would turn to the right (clockwise) in response to this effect. In the Southern Hemisphere the effect is reversed.

The shape of the continents, global wind patterns, and the Coriolis effect combine to form the massive surface currents. The Gulf Stream is one of the most powerful and influential currents of the earth and typifies a western boundary current. These strong, fast, and deep currents are not extraordinarily wide, but they carry tremendous amounts of warm tropical water toward the poles in most oceans. The salinity of water in these currents tends to be high and the edges clearly defined.

Eastern boundary currents formed at an ocean's eastern edge have the opposite characteristics from those of their western counterparts. The water in eastern boundary currents, coming from high latitudes, is cold, relatively low in salinity, shallow, and broad. The edges of the eastern boundary currents are not clearly defined, and the speed of these currents is typically much less than that of a western boundary current.

Both eastern and western boundary currents have important effects on weather. England is in the path of the warm Gulf Stream (via the North Atlantic Current), and because of this experiences weather much milder than its northerly latitude would lead one to expect. Likewise, the weather that led Mark Twain to observe that the coldest winter he ever spent was a summer in San Francisco is governed by the close passage near that city of the cold California Current. The fogs, low summer daytime temperatures, and other weather effects in San Francisco are similar to those expected of a city farther north.

Gyres in the ocean are completed by lateral (east–west) currents that transport water from these eastern and western boundary currents to the other side of the ocean to close the circle. The trade winds near the equator and the westerlies at around 45° N and 45° S provide the power for this movement. It is on these currents that fishnet floats travel from the water near Japan to the beaches of California, and on which Thor Heyerdahl's *Kon Tiki* drifted from South America to Polynesia.

Not all ocean currents flow at the surface. One of the most unusual is the Cromwell Current, a subsurface current at the equator running from east to west. Much deeper currents have been observed, including one below the Gulf Stream, which flows in the opposite direction at depths up to 2,730 m (9,000 ft). Another current at the abyssal floor in the central Pacific moves at the stately pace of a few hundred feet a year. These deepest of currents form when chilled, salty, dense water sinks during the freezing of polar pack ice. The identifiable water masses slip to the bottom along the continental shelf and creep horizontally along the seabed for great distances. *Antarctic Bottom Water*, densest of all seawater, has been identified at 50° N latitude near Alaska's Aleutian Islands!

Not all currents flow horizontally. Wind blowing from the north along the West Coast of the United States can cause water at the surface to move to the west, offshore (remember the Coriolis effect). This water is then replaced by cold, nutrient-rich bottom water in a process known as upwelling. A plankton bloom often occurs following this movement.

Alterations in the normal flow of surface currents can also cause a catastrophic decline in plankton productivity. A phenomenon known as El Niño along the west coast of South America occasionally interrupts the northward flow of the cold and rich Humboldt Current from the south. When the normal flow of cold nutrient-laden water is replaced by a back-flow of warm depleted water from the equatorial regions, the oceanic productivity of the region plummets. The winter of 1972–73 was a time of El Niño. During this period Peru dropped from number one in fish caught per capita (1971) to number six (1974), largely due to the effects of El Niño.

The 1982–83 El Niño was the most spectacular El Niño event of this century. Starting in July of 1982, the southeast trade winds that normally blow away from the west coast of South America slowed over the ocean north and south of the equator. The great gyres of the Pacific, deprived of one of their most important driving winds, began to move sluggishly. With movement of the cold Humboldt Current waters into the tropical areas slowing down, equatorial ocean temperatures rose. The westward flow of water away from Central America at the equator stopped and actually reversed. Hot water "backed up" to the north and south, causing warmer waters off our own Pacific Coast and off the coast of southern Peru. Sea levels rose, becoming 10 cm (4 in.) higher than normal at San Diego. (Consider the volume of water necessary to make that magnitude of change!) The increase in offshore temperatures injected more water vapor into the atmosphere off the west coast of the American continents, and catastrophic rains ensued. Ecuador and Peru experienced serious flooding, and the west coast of North America saw many storm and rainfall records broken. Many of you will remember the television news reports of high surf shattering piers and coastal homes in southern California during that first El Niño winter.

Instead of dwindling in a normal fashion, this El Niño continued for an uncharacteristically long time. Two years after the event, the ocean off California was still warmer than usual, making life difficult for some local temperature-sensitive species including some varieties of kelp and fish. The unusual variety of exotic fish attracted by the warmer water, however, were a delight to sport fishermen. Studies have shown the 1982–83 event affected the Atlantic as well. A smaller El Niño event began in the Pacific in January of 1991, and disappeared in the fall of 1992. Computer modeling has been used to study these and other related occurrences, now called ENSO events (*El Niño – Southern Oscillation*). Some hope that improved weather forecasting will result from lessons learned through the last two ENSO cycles.

Oceanographers study currents by various means, ranging from the use of thousands of floating bottles to sophisticated research submarines. Deep currents are analyzed by automatic recording current meters placed on the ocean floor. Currents in middle depths have been surveyed by parachute drogues with pingers attached to assist in tracking them as they drift. Surface currents may be investigated by a variety of means, but foremost among these in recent years has been the research submersible, appropriately named *Benjamin Franklin*. This vessel was designed to drift for weeks at a time at a set depth within the Gulf Stream to allow scientists ample opportunity to sample the physical and biological characteristics of the surrounding water. New satellites and sensors being readied to fly on future space shuttle missions should greatly advance the study of ocean currents.

Weather, climate, and food are all affected by ocean currents. Currents do considerably more on this planet than transport messages in bottles.

Learning Objectives

After completing the reading assignment and viewing the program, the student will be able to:

☐ List the physical factors that cause ocean currents.

☐ Understand the general circulation pattern of surface currents.

☐ Contrast western boundary currents with eastern boundary currents.

☐ Explain some of the effects of currents on surface productivity and localized weather conditions.

☐ Describe some characteristics and causes of deep currents and vertical currents.

☐ Describe an El Niño effect in the Pacific.

Key Terms and Phrases

Current A mass-flow movement of water.

Geostrophic current A large, horizontal movement of water caused by major physical factors. These factors include wind, the Coriolis effect, position and size of continents, ridges, islands, and so on. These currents are literally turned by the earth, and they form a whirlpool-like gyre within an ocean.

Gyre A set of geostrophic currents (usually four) within an ocean. Gyres in the Northern Hemisphere turn clockwise and in the Southern Hemisphere they turn counterclockwise.

Coriolis effect An apparent force of moving particles resulting from the earth's rotation. The Coriolis effect causes moving particles to be deflected to the right of motion in the Northern Hemisphere and to the left in the Southern Hemisphere. The force is proportional to the speed and latitude of the moving particle and cannot change the speed of the particle. (From text Glossary)

Upwelling Vertical movement of water in an upward direction. Usually occurs along a coast.

Eastern boundary current An ocean current along the eastern boundary of an ocean (offshore from a western continental margin). Example: California Current.

Western boundary current An ocean current along the western boundary of an ocean (offshore from an eastern continental margin). Example: Gulf Stream.

El Niño A general disruption of the equatorial flow of Pacific Ocean water away from the coast of South America thought to be caused by a cessation or slowing of the equatorial trade winds. The results of an El Niño can include a local rise in sea level, increased sea surface temperatures in the eastern Pacific area, thermal stress on organisms, increased coastal rainfall, and other associated phenomena.

Special Note: When discussing currents and winds, we face potential confusion, so we need to clarify a number of expressions now.

Winds are named by where they come from. A north wind, for example, comes from the north and in our hemisphere is often cool or cold. A west wind comes from the west. On the other hand, ocean currents are named by where they are *going*. A northern current goes to the north. A western current moves from east to west. An exception is the West Wind Drift, a current north of the Antarctic continent, that moves toward the east. The reason for this discrepancy is that the current is itself named after the wind that drives it, which comes from the west.

Before Viewing

☐ Read Chapter 9, "Ocean Circulation," pages 208–236 in Garrison; or Chapter 10, "Ocean Circulation," pages 150–180 in Ingmanson & Wallace.

☐ Watch Program 12: "Ocean Currents."

After Viewing

Reread the text sections, being sure you understand the key terms and phrases, and complete the following exercises:

1. If moving objects in the Northern Hemisphere veer off course to the right, why do the currents seem to move for great distances over seemingly straight, great circles? Look at Figure 9.8 on page 213 in Garrison or Figure 10.2 in Ingmanson & Wallace, and note the directly westward path of the North Equatorial Current, for example. It doesn't veer northward until it reaches the vicinity of the Philippines. Why not?

2. If western boundary currents are the strongest currents in the Northern Hemisphere, are eastern boundary currents the strongest in the Southern Hemisphere? After all, doesn't the Coriolis effect cause reversed flow directions for geostrophic currents there?

3. Are there any major ocean currents that are not geostrophic?

4. Are there any ocean currents that are never deflected by land?

5. What causes El Niño?

Answer Key for Exercises

1. To understand the answer to this question you will need to look at Figures 9.6 and 9.7 on pages 211 and 212 in Garrison or Figure 10.9b in Ingmanson & Wallace. Because of the large mass of water moving clockwise around the perimeter of the Pacific Ocean, and because of the rotation of the earth and the operation of the Coriolis effect, there is actually a bulge of water near the center of the ocean. The bulge is only about 2–3 m (6–9 ft) high, and is offset to the west of the geometric center of the Pacific. The bulge is not an abrupt hump, but is instead a smooth progression from ocean edge to center to ocean edge. Water moving toward the west in the North Equatorial Current would like to obey the Coriolis urge to turn to the right (north). But it cannot, because to do so would mean moving "uphill." So gravity and the Coriolis force offset each other, and the water continues to move westward until turned north by the intrusion of a continental mass.

   A moment's reflection will reveal something of a "chicken/egg" problem. Which came first, the bulge or the deflection? At some point in the distant past, as continents moved toward their present positions, water must have moved toward the ocean's center, thereby causing the bulge. Since then the water flowing in the geostrophic current system around the Pacific has behaved as it does today.

2. Careful here! The earth turns toward the east in both hemispheres. Currents in both hemispheres move toward the west near the equator, turn poleward as they approach land, and then flow eastward near the poles and toward the equator along the eastern boundaries of the oceans. Think about it. One of the factors that makes western boundary currents so concentrated is a sort of planetary "crack-the-whip" effect. The westward-flowing currents are intercepted by the eastward-moving continents and flung poleward. So, western boundary currents are strongest in both hemispheres. An exception is the current off the east coast of Australia. Landmass configuration and the many small islands there do not permit strong, concentrated water movement.

3. Yes, many of them. The currents at the mouths of rivers, tidal currents in harbors, wind-induced vertical currents, and others. Geostrophic currents are those major currents within an ocean that form its perimeter and are influenced by winds, the Coriolis effect, and the shapes of ocean basins.

4. Yes. The West Wind Drift north of the continent of Antarctica is such a current. Because of the configuration of winds and ocean bottom, the currents running there never directly encounter land. Look at the West Wind Drift at the bottom of Figure 9.8 in Garrison or Figure 10.2 in Ingmanson & Wallace.

5. The easy answer is "the trade winds stop." But the really difficult question then becomes: "Why do the trade winds stop?" And that one is a stumper. Many hypotheses have been advanced to explain the El Niño, but so far none has been agreed upon by oceanographers and meteorologists. Some of these hypotheses are based upon solar constant variations (differences in the amount of solar energy put out by our sun); the number and severity of tropical cyclones in an area (which could affect the heat balance of the ocean in the area of the storms); and even the pollution of the atmosphere by natural and man-made effects. So far, though, no one is sure. You can be sure, however, that much research is going in that direction.

Optional Activity

Trite though it may sound, putting messages in bottles and tossing the bottles into the ocean can be very instructive. As your text mentions, Woods Hole Oceanographic Institution has been doing just that for years. The best bottles are heavy-walled beverage bottles that can be corked. The message inside should ideally be in several languages, and if a reward is offered, you should certainly uphold your end of the bargain when the message is returned. Seal the bottles with hot wax over the cork and set them free at a distance from land. With luck and a good chart, you should be able to learn something about the currents in the area of release.

Self-Test

1. Benjamin Franklin was the first person to recognize the extent of the _____ and to publish a navigational chart of it.
   a. North Equatorial Current
   b. Gulf Stream
   c. El Niño
   d. uroshio Current
   e. Canary Current

2. Geostrophic currents can be described by which of the following statements?
   a. They are turned by the earth.
   b. They depend on the Coriolis effect and the position of continental landmasses for their direction.
   c. They form gyres around ocean perimeters.
   d. They are physically large.
   e. All of the above are true.

3. Which of these statements is true?
   a. The Coriolis effect is strongest at the equator.
   b. The Coriolis effect has no real effect. It is only an imaginary force on the earth.
   c. The Coriolis effect causes moving objects to deflect to the right (clockwise) in the Northern Hemisphere and to the left (counterclockwise) in the Southern Hemisphere.
   d. The Coriolis effect causes moving objects to deflect to the left (counterclockwise) in the Northern Hemisphere and to the right (clockwise) in the Southern Hemisphere.

4. The earth's strongest currents are
   a. western boundary currents
   b. equatorial currents
   c. eastern boundary currents
   d. rivers
   e. tidal currents

5. Which current within a Northern Hemisphere gyre would you expect to have the lowest salinity and temperature?
   a. western boundary current
   b. eastern boundary current
   c. southern boundary current
   d. northern boundary current
   e. All of these would be about equal in temperature and salinity.

6. A gyre may best be defined as
   a. a whirlpool
   b. an interlocking set of (usually) four geostrophic currents within an ocean basin
   c. a swirl of water turning always clockwise
   d. a swirl of water turning always counter-clockwise
   e. none of the above

7. Deep currents generally move _____ than the surface current above, and in _____ direction.
   a. faster . . . the same
   b. faster . . . the opposite
   c. slower . . . the same
   d. slower . . . the opposite
   e. at the same speed as . . . the same

8. The current that never turns from course because of land is the
   a. West Wind Drift
   b. Canary Current
   c. Humboldt Current
   d. Arctic Current
   e. Kuroshio Current

9. El Niño results due to an interruption of flow of the _____ Current(s).
   a. Kuroshio
   b. Humboldt and Equatorial
   c. Canary and North Atlantic
   d. West Wind Drift
   e. Arctic

10. Which of these is generally *not* an El Niño effect?
    a. a rise in the sea level off the coast of the American continents
    b. an increase in sea temperature in the eastern Pacific
    c. increased rainfall in west coastal countries or states of the American continents
    d. an often catastrophic decrease in the commercial fisheries of the affected countries
    e. a decline in exotic species of fish and plankton in the affected waters

Supplemental Reading

Graham, N., and N. White, "The El Niño Cycle: A Natural Oscillator of the Pacific Ocean-Atmosphere System." *Science 240*, vol. 240 (1988). (Excellent though technical article summarizing recent ENSO research.)

Idyll, C. P., "The Anchovy Crisis." *Scientific American* (1973).

Knauss, J. A., "The Cromwell Currents." *Scientific American* (1961).

McDonald, J. E., "The Coriolis Effect." *Scientific American* (1952).

Oceanus, vol. 27, no. 2 (1984). (This issue is largely devoted to an explanation of the causes and effects of the 1982–83 El Niño event. Highly recommended reading, *Oceanus* is available in most good libraries or directly from the Woods Hole Oceanographic Institution in Woods Hole, Massachusetts.)

Stommel, H., *A Discussion between a Chief Engineer and an Oceanographer about the Machinery of the Ocean Circulation*. 1988. Princeton: Princeton University Press. (Excellent book for the interested general reader.)

Stommel, H., "The Circulation of the Abyss." *Scientific American* (1958).

Spinel, R. C., and P. F. Worcester, "Ocean Acoustic Tomography." *Scientific American*, Oct. 1990, 94–99. (Recent information on the predominance of midscale circulation.)

Lesson Thirteen   Wind Waves and Water Dynamics

Overview

Watching waves is a fascinating pastime, whether at the edge of a pond, on a steep beach, or hanging over the rail of a pitching ship. The variety of waves seems endless, from the ripples that flicker as a breeze wafts by, to the mountainous torn seas and howling winds that threaten the survival of the unlucky mariner. These are the extremes in a broad spectrum of wave motions that continually upset the equilibrium of the sea surface.

Sea surface wind waves are not the only types of wave motion in the oceans. Sounds propagated within the sea are also waves, generated by different sources – biological and man-made – and traveling great distances. Internal waves are generated within the sea between layers of different water densities whose actions seem unrelated to the surface wind waves. The tides are a form of wave resulting from the gravitational attraction of the moon and sun. The greatest waves are the tsunami or seismic sea waves that deceptively race across the oceans as long, low waves only to lift at the shore and plunge inland as a crashing wall of water.

In spite of differences in origin and size, waves have certain fundamental properties in common. They can all be described using similar parameters: wavelength, height, amplitude, period, and velocity. In every case in the open sea, the wave is transmitting energy through the water. Little of the water itself is carried along with the wave form.

The surface of the sea is never still, as the devoted wave watcher knows. Even on a calm summer day breakers can be considerable, for the waves that break on the shore are more than just coastal phenomena. They are survivors of a storm or disturbance that may be half an earth away. West Coast surfers enjoying summers at the beach may be riding waves that started two weeks earlier in winter Antarctic storms down in the Roaring Forties. Although the surfer is not usually concerned with where the waves originated, or how the wind imparts energy to the water, serious oceanographers are. Even with advanced mathematics, wave theory, wave tanks, and modern technologies, oceanographers are still hard pressed to explain the complexities of actual waves at sea.

The first grip of wind on the water surface results in little capillary waves that form when the wind velocity is only a breezy half-knot. Once the surface becomes irregular, the wind is more effective at imparting energy to the water, and the waves grow. The ultimate height waves will reach depends on the force of the wind, the length of time the wind continues to blow, and the distance (fetch) the waves move out under the wind. The long, hard "blows" over a wide storm area can create wind waves that have been measured at heights over 30 m (100 ft). Most waves, however, are usually under 3 m (10 ft).

Once started, the wave form moves out in all directions from the storm center. The smaller, slower waves tend to be overtaken by the longer, faster waves. The sea surface gets sorted out, and the choppy confused sea becomes a surface of rounded hills and valleys called the swell. In the deep water, waves are not affected by the ocean bottom. Their velocity is dependent on the wave-generating force of the wind. No matter how high and fierce these waves might be, they do not move the deep waters of the sea, and pass unnoticed at depths greater than half their wavelength. Since wavelengths rarely exceed 152 m (500 ft), most wind waves stir the water only in the upper 76 m (250 ft) or less.

When the swell approaches the shore and the water becomes shallow, the waves now "feel bottom"; and friction between the moving water and the seafloor becomes important. Remarkable changes take place: The wavelengths shorten and the wave crests tend to bunch up. The wave heights increase, and the velocity slows down. The wave form peaks, becomes unstable, the top falls over and the wave breaks. It is not unusual to be offshore on a boat that is barely rocking in a gentle swell and watch with astonishment as the waves lift near shore and form huge breakers that crash on the beach.

Occasionally a group of high waves may contain a solitary giant "rogue wave," which rises suddenly out of the sea. The chances of encountering such waves are slim but have increased in recent years as shipping has expanded over the seas. Large ships seem more vulnerable, and even the venerable *Queen Mary* almost capsized when hit by a single mountainous wave off the coast of Scotland. The origin of these waves is probably related to special conditions of wave interference or interaction between currents and swells.

Not all waves are at the sea surface. *Internal waves* are more common than surface waves but are less obvious because they occur within the water column at boundaries between layers of different densities. Measurements of temperature and density have shown that whole sections of the ocean are moving up and down as the internal waves move in all directions. The sea is never still. These waves can be very large with an amplitude of perhaps 30 m (100 ft) or more, with wavelengths of about a half mile, and a period of 5 to 8 min. They have been detected in the famous Loch Ness and in other lakes, harbors, and bays, as well as in the deep sea. The only surface evidence of internal waves might be long lines of smooth water or "slicks." They are apparently generated by tides, changes in atmospheric pressure, or any process that creates surface waves. Internal waves are of interest to scientists and the military because the density changes they produce can affect the buoyancy and pressure on submarines.

Another wave found in bays, harbors, and lakes is the *seiche*, a standing wave that does not progress forward, but instead rocks or sloshes back and forth. Many harbors exhibit this phenomenon, but the height of the seiche is so low that the wave passes unseen. Ships, however, may suddenly rock or pull on their mooring lines, even in the absence of wind or wind waves, sometimes damaging docks, piers, and other facilities.

The largest and most fearsome waves are the *tsunami,* or seismic sea waves (incorrectly called tidal waves), set in motion by movements of the earth's crust. Any bouncing of the seafloor, coastal earthquakes, underwater landslides, or volcanic eruptions at sea may generate these long-period, swiftly moving waves that, from time to time, move out across the oceans. In the open ocean, tsunami may reach a velocity of 760 km/hr (471 mph) while the height of the wave may be no more than 1 m (3 ft). The wavelength is so long, about 150 mi, that the wave feels bottom even at the greatest depths. In fact, the sea cucumber on the floor of the abyssal plains may sense the tug of a passing tsunami, while the passengers lolling on the deck of a cruise ship above may be totally unaware of the energy roiling the sea beneath them.

One of the peculiarities of the tsunami is that when the wave finally reaches a shore, the trough comes in first and the water drains seaward, frequently exposing the floor of the bay or harbor. Unknowing people have run down to the beach to pick up crustaceans, flopping fish, and other stranded sea creatures, only to be inundated a few minutes later when the crest of the 50-ft wave comes roaring in. Other high crests may come in, fifteen or so minutes apart, before the wave system dissipates and the sea surface settles into the normal wave patterns.

Today there is an elaborate tsunami warning service centered in Hawaii that monitors all earthquakes and volcanoes in the Pacific Basin. The danger is still present and in some ways increasing because of coastal urban development on the mainland and on many islands. But the tsunami warning service has already saved many lives and is a fine example of practical oceanography at work.

Learning Objectives

After completing the reading assignment and viewing the program, the student will be able to:

☐ Understand the movement of water particles in various kinds of waves.

☐ Define each of the parameters used to describe waves.

☐ Describe how wind waves are generated and move in the open ocean.

☐ Discuss wind waves at the shore, including types of breakers, rip currents, and the processes of refraction, diffraction, and reflection.

☐ Compare tsunami as to origin, properties, and effects with normal wind waves.

☐ Describe special waves such as seiches and internal waves and rogue waves.

Key Terms and Phrases

Wavelength The length of one wave measured from crest to crest. Wind waves will average from about 60 or 70 m to about 150 m in the open sea.

Wave period The time it takes for one wavelength or wave to pass a given point; usually expressed in seconds. The wave period usually varies from about 4 to 10 sec, but it can be from less than 0.1 sec to 24 hr.

Propagation rate or wave velocity Wave speed. In shallow water the propagation rate depends on depth since passing waves feel bottom. In deep water it depends on the energy imparted to the wave form and wavelength. Longer waves travel faster.

Note: Other terms that apply to anatomy of waves are defined on pages 181–182 and in Figure 11.1 of Ingmanson & Wallace, or on pages 228–229 and in Figure 10.2 on page 239 of Garrison.

Capillary waves The first ripples to form when the wind blows. The wave period will be less than 0.1 sec, and the height and wavelength will be very small. Capillary waves are dominated or torn down by the surface tension of the water.

Gravity or wind waves Most of the waves in the open ocean. The wave period ranges from less than 1 sec to a theoretical 30 sec. Height may vary from a few centimeters to 10 to 12 m. The highest wave ever measured in the open sea exceeded 30 m! Gravity or wind waves are dominated by gravity that restores the sea surface to a flat equilibrium position.

Motion of water particles Water particles in a passing wave move in circular orbits, moving forward as the crest passes and backward at the next trough. The diameter of the orbit is equal to the wave height at the surface. The orbits get smaller with depth, and at depths equal to one-half the wavelength, the water motion is minimal. In shallow water, the orbits of the water particles become elliptical, and the water interacts with the seafloor or feels bottom.

Mass transport The small forward motion of water in a passing wave, a factor negligible in the open ocean since only the wave form moves. Mass transport may be important at the shore, especially during storm surges.

Rip current A localized, narrow, seaward-moving current in the surf zone, usually in a channel or gap in a sand bar, incorrectly called a rip tide or undertow. Rip currents are dangerous to swimmers because of their high velocity, about 3 knots.

Wave refraction The bending of the wave front as it approaches the shore. When a wave approaches the shore at an angle, part of the wave will be in shallow water and will slow down; part will still be in deep water and will move ahead. The net result is that waves tend to approach the coast almost parallel to the shore.

Tsunami Seismic sea wave incorrectly called a tidal wave. Tsunami is a long-period wave of 15–20 min with a high velocity of 760 km/hr (over 471 mph). In the open ocean a tsunami is a low wave of less than 1 m in height. At the shore, it may rise over 30 m. Generated by landslides, movements of the seafloor, volcanic explosions, and earthquakes, tsunami are most common in the Pacific Basin.

Storm surges High water level as a result of large storms, high tides, and high winds associated with low pressure hurricanes. Storm surges may lift water level over 6 m and may be very costly to life and property.

Internal waves Waves that occur within the sea at layers of different water densities. The heights of internal waves may vary from 4 to over 30 m; periods from about 5 to 8 min; wavelength from 600 to 900 m. They were discovered by oscillations in temperature and density at depth and may be recognized at the sea surface by long lines of smooth water or slicks.

Before Viewing

☐ Read Chapter 10 *and* pages 259–267 in Garrison, or Chapter 11 in Ingmanson & Wallace. Note particularly the diagrams throughout these chapters.

☐ If you enjoy a little math, see pages 241–242 in Garrison or pages 181–182 in Ingmanson & Wallace. Pay careful attention to the speed of tsunami – these great waves travel at the speed of a jetliner!

☐ Many words in the Glossary will aid in understanding wave phenomena. Check the following:

capillary wave	storm surge
dead water	surf
deep-water wave	surface wave
diffraction	swell
fetch	tsunami
gravity wave	wave
internal wave	wave crest
ocean slick; slick	wave height
reflection	wavelength
refraction	wave train
seiche	wave trough
shallow-water wave	

☐ Watch Program 13: "Wind Waves and Water Dynamics."

After Viewing

Reread the text sections, being sure you understand the key terms and phrases, and complete the following exercises:

1. Describe the motion of water particles in a wave passing over deep water, then passing over shallow water. See Figure 10.6 on page 242 of Garrison, or Figure 11.2 on page 182 of Ingmanson & Wallace. Why is mass transport of little importance in the open ocean, but of considerable effect during a storm surge on a flat-lying coast?

2. You are sitting on a beach and a large wave crashes on the shore. Where did it all start? What happened on the way? What changes occurred in the swell as it approached the shore? Describe the forces required to build a wave to its full size.

3. If one story on a building is about 10 ft, how many stories high was the wave measured by Lt. Commander Whitemarsh of the USS *Ramapo*? See page 246 of Garrison, or page 191 of Ingmanson & Wallace.

4. You are on a vacation in Hawaii. What kind of waves are you likely to encounter, spillers or plungers? Is the slope of the bottom steep or gentle? Good or poor surfing? Some of the best surfing is off the north coast of Oahu in the Hawaiian Islands. What factors might contribute to the good surfing there?

5. You are caught in a rip current! Describe what is happening to you and how you might get out of it.

6. Describe the causes of a tsunami. If you were on a ship at sea, what would you see as a tsunami passed? If you were in Half Moon Bay, Alaska on April 1, 1946, what would you have seen? What is being done to prevent loss of life from tsunami?

7. The waters of bays and harbors sometimes rock like the water in a bathtub. What do oceanographers call these standing waves, and why are they a problem? What might be some of the causes? See page 263 of Garrison, or pages 202–203 of Ingmanson & Wallace. By the way, the correct pronunciation of the word *seiche* is "saysh."

8. Why are internal waves of great concern to operators of submarines? At any one depth, what changes in the properties of the water might occur with the passing of an internal wave?

Answer Key for Exercises

4. The surfing is particularly good on the north coast of Oahu because the waves are generated from storms in the North Pacific near the Aleutians. They have a long fetch, building as they travel under the wind, and have a chance to reach their full height. There are no islands or land masses between the north coast and the Aleutians to impede their progress.

8. Internal waves were discovered by the wide variations in temperature and density encountered while measuring deep-sea

temperatures. The temperatures cyclically changed as the cold water moved up and down within the passing internal wave. Apparently the oceans are in constant motion internally, as various internal waves move in different directions.

Optional Activities

1. When you are at the beach, count the seconds between breaking waves. This will give you the wave period. If the period turns out to be 15 min, what should you do? RUN!! It's a TSUNAMI! What else might be happening that would tell you a tsunami is on its way?

2. Locate the Red Sea on the map at the back of the text. What might have happened there if the waters parted long enough to allow Moses to lead his people out of Egypt, then inundated the pursuing Egyptians?

3. Study the Pacific Basin on the map at the back of the text. According to plate tectonic theory, are the margins active or passive? Does the theory explain why most of the tsunami occur in the Pacific? Locate Krakatoa in the Indian Ocean, west of Java, the site of one of the greatest explosions in historic times. Over 36,000 people perished although very few actually lived on Krakatoa. How would you explain this?

4. The Pacific Basin is surrounded by the Ring of Fire, a belt of earthquakes and volcanoes. Tsunami, as we know, can be generated by either. Note in Figure 11.5 on page 264 of Garrison, or Figure 11.22 on page 199 of Ingmanson & Wallace, that wherever a tsunami is started, whether in Chile, Middle America, Alaska, Japan, or the Philippines, the spreading disturbance will hit Hawaii. When there is an earthquake that might start a tsunami, the information is immediately sent to the Information Center in Hawaii, evaluated, and sent to all areas that might be affected. From this diagram, can you determine how many hours it took for the Good Friday tsunami from Alaska to reach Hawaii?

Answer Key for Optional Activities

3. The tsunami that followed the eruption of Krakatoa rose to heights of over 30 m (almost 100 ft) and inundated the many small but densely populated islands and low-lying coasts of Indonesia. The energy from the wave crossed the Indian Ocean, passed the southern tip of Africa, and moved north in the Atlantic Ocean into the English Channel where it caused a very small rise in sea level.

Self-Test

1. Which statement best describes water particles in a deep-water wave?
   a. They move rapidly toward the shore.
   b. They move in circular orbits.
   c. They move in elliptical orbits.
   d. They move up and down in one place.
   e. They do not move at all.

2. Mass transport is an important physical phenomenon
   a. in the open sea
   b. in the motion of capillary waves
   c. when the wind is at about one knot
   d. on low-lying coasts during hurricanes and strong on-shore winds
   e. all of these

3. In order to develop large waves, which of the following conditions are necessary?
   a. high velocity winds
   b. long fetch
   c. storms over a wide area
   d. storms persistent over a long period of time
   e. all of these

4. The first grip of the wind on the water results in
   a. a swell
   b. gravity waves
   c. seiches
   d. capillary waves
   e. spillers

5. Which of the following statements best describes long-period waves?
   a. They have a greater velocity.
   b. They have a shorter wavelength.
   c. They do not move the water to very great depths.
   d. They are seen as capillary waves or "chop" on the surface.
   e. All of the above are true.

6. In the open ocean, the surface waves will affect the water
   a. in the surface orbits of the water particles only
   b. to a depth equal to one-half the wavelength only
   c. to a depth equal to one-half the wave velocity only
   d. from top to bottom, if the wavelength is less than 150 m
   e. none of these

7. Waves at the shore will break
   a. when the wave starts to feel bottom
   b. when the wave form changes from rounded to a peak
   c. when the internal angle at the peak becomes less than 120°
   d. when the water depth is four-thirds the wave height
   e. all of these

8. Tsunami can be generated by
   a. underwater or coastal landslides
   b. coastal earthquakes
   c. volcanic eruptions
   d. all of the above
   e. none of the above

9. Which of the following is *not* characteristic of tsunami in the *open sea*?
   a. velocity approaching 760 km/hr (471 mph)
   b. wave period of 15 to 20 min
   c. height in the open ocean up to 30 m
   d. wavelengths over 240 km (150 mi)
   e. groups of three or four waves together

10. When waves move out of a storm area
    a. the faster waves overtake the slower waves
    b. the longer waves take in the shorter waves
    c. the waves sort out into packets called wave trains
    d. the sea surface smooths out into the swell
    e. all of the above

11. Long, smooth, parallel marks on the sea surface are considered evidence of
    a. long-period surface waves
    b. standing waves
    c. internal waves
    d. undertow
    e. approaching plunging waves

12. Waves that are not obvious on the surface of the water, but that move in all directions through the oceans, lakes, and harbors, are
    a. seiches
    b. internal waves
    c. rogue waves
    d. capillary waves
    e. gravity waves

Supplemental Reading

Kampion, D., *The Book of Waves*. Santa Barbara, CA: Arpel Graphics. 1989. (The definitive coffee-table book of waves. Magnificent photography.)

Lesson Fourteen   The Ebb and Flow

Overview

The rhythmic rise and fall of the tides is one of the more familiar daily events in coastal ocean-ography. The intertidal organisms, as we have seen, are totally tuned in to the tidal cycle, whether at the shore or in some laboratory tank out of sight and sound of the sea. Beach people, dam diggers, and coastal navigators adjust their life activities by the tides, but seldom stop to consider the forces that cause these pulsations of the ocean, of lakes and ponds, of the atmosphere, and of the solid crust of the earth itself.

The Greeks and Romans gave detailed accounts of tidal phenomena which they correctly perceived as influences of the moon and sun. It wasn't until the seventeenth century, however, that Sir Isaac Newton published the theory of universal gravitation that laid the groundwork for tidal analysis. Today tidal theory has entered a new phase with the use of computers. Although electronic calculations are remarkably accurate, certain tidal movements are still not fully understood. For example, we cannot make exact tidal predictions for bays or harbors where no previous records have been kept.

As Newton demonstrated, the deceptively gentle play of the tides on most beaches is an event of cosmic proportions. The force that holds the planets in their orbits around the sun and holds the sun and its 2 billion companion stars in the galaxy, and that flows throughout the universe is gravitation, the same force that moves the waters of the earth. In the case of the earth's tides, the major force is exerted by the combined gravitational pull of the moon and sun. As Newton's equations show, the force of gravitation between two bodies is the product of their respective masses divided by the square of the distance between them. The larger the two

bodies, the greater the attraction. But as the distance between the masses increases, the force between them decreases rapidly. For example, the mass of the sun is about 27 million times that of the moon, but the moon is about 387 times closer to Earth and exerts a tidal pull about twice that of the sun.

On the side of the earth facing the moon, the moon's gravity pulls the water into a high-tide bulge. On the side opposite, another bulge appears, which is related to the movement of the earth about the center of the earth-moon system. Because the moon also exerts a slight pull on the earth, the waters on the far side are left to lag and rise. While areas under these bulges are having high tides, the seas between are experiencing low tides. As the earth rotates on its axis, the waters are held by the sun and moon, and the land and oceans move into the high- and low-tide regions with each section of coast having two highs and two lows each tidal day.

The tides that occur at any one place on Earth are the result of at least 150 interacting forces. You can see why tide prediction is still something of an art, rather than a completely predictable science. Some of the factors that affect tidal analysis are: the revolution of the moon about the earth in approximately 28½ days; and the elliptical orbit of the moon varying from the near point, *perigee*, about 220,000 mi, to the far point, *apogee*, about 248,000 mi. (Remember that gravitational pull is strongly affected by distance.) Also, the plane of the moon's orbit is tilted relative to the earth's orbit by about 5°, a factor known as "declination." The earth's orbit about the sun is also elliptical, carrying the earth from the near point, *perihelion*, in January (about 91,400,000 mi) to the far point, *aphelion*, in July (about 94,800,000 mi).

Tides change in regular cycles following the phases of the moon. During the full moon when the moon is on the side of the earth away from the sun, *spring tides* occur. These tides have the greatest tidal range – the highest highs and the lowest lows – because the sun, earth, and moon are in a straight line and gravitational pull is at a maximum. Again, at new moon when the moon is directly between the sun and earth, spring tides will occur. When the moon is at first or third quarter and pulling at right angles to the sun, the tidal range will be at a minimum, producing the lowest of the high tides and the highest of the low tides.

To predict what the tides will be, the characteristics of the ocean basins, the continental shelves and coast, and even the properties of the water must be considered. Tides are actually very long waves that move the whole water column down to the seafloor. The irregular topography of the ocean – including the midocean ridges and rises, the islands and seamounts, the deep trenches – affects the flow of waters. The water itself varies in density because of the differences in temperature, pressure, and salinity. This also affects the character of the flow. Even the Coriolis effect must be factored into a complete tidal analysis.

The actual height and range of a local tide still depends on many other conditions, such as the shape, size, and depth of the bay or harbor; the natural rocking motion of the harbor; and the nature of the opening to the harbor, estuary, sound, or bay. A broad continental shelf tends to dissipate most of the tidal energy by friction, sometimes reducing the range to inches.

The sum of the tide-generating forces results in wide variations in tidal range and tidal patterns. For example, in the United States the East Coast has two high and two low tides of equal height per lunar day. The West Coast also has two highs and two lows, but of different heights. The Gulf of Mexico has only one high and one low tide per lunar day. The variation in tidal range from highest high to lowest low is extreme. The highest tides in the world are in the Bay of Fundy, between Nova Scotia and New Brunswick, where the range may be well over 15 m (50 ft). Some areas such as the Mediterranean, the Gulf of Mexico, and the Baltic Sea have almost no tidal variances. Tahiti, Hawaii, and other islands of the open Pacific have a tidal range of no more than 0.6 m (1–2 ft). Off the coast of France where the shelf is very broad and flat, as at the famous Mont-Saint-Michel, a moderately high tide can bring the sea in almost 10 mi.

A chart of all the tides existing in the world ocean at any moment reveals a picture far more complex than one might expect from the simple gravitational effects of the moon and sun. Instead of a pair of tidal bulges advancing westward under the moon, there are perhaps a dozen discrete cells, called amphidromic points, where the tidal range is zero at the center increasing outward. The tides turn about these points in a manner similar to the waves created when a circular pan is rotated and rinsed, or when gold is panned. The line of high tide intersects the coast at approximately right angles. Along the West Coast of the United States, the high-tide wave moves north from the tip of Baja California, past California and Oregon, to Washington at an average velocity of about 640 km/hr (400 mph).

When the high tide intersects a narrow opening to a bay or harbor, it may set up a swiftly flowing tidal current that floods inland, and then ebbs seaward at low tide. One of the most exciting phenomena is the tidal bore, a wall of water that rushes upstream in certain rivers and funnel-shaped bays. When the rising tide encounters the outflowing waters of the river, the force of their meeting creates a surging, turbulent wave that may surge upstream at over ten knots. Great whirlpools sometimes develop, the most famous of which is the maelström off the west coast of Norway, graphically described by Edgar Allan Poe's luckless mariner. If you are using the Garrison text, read the box on the maelström on pages 282–283.

Tidal power has long been a dream that would free us from our dependency upon fossil fuels and nuclear power plants. In spite of the fact that tides are a clean, safe, reliable, and renewable form of energy, few plants have ever been built.

The most effective way to generate electricity from tides would be to build a dam across part of a bay that has a narrow opening and a large tidal range. At high tide a basin or reservoir behind the dam would fill with water. At low tide, the water would be released through turbines. Existing and proposed tidal power facilities in Canada, France, South Korea, and the USSR utilize this method of producing electricity. A proposed project in the Bay of Fundy is estimated to be able to produce the equivalent of the output of 250 large nuclear power plants. Yet the costs are high and tidal power is not economically competitive. Also, the environmental impact of tidal dams and installations will have to be considered carefully. Because tidal power could eventually make a considerable contribution to the global energy needs, it should remain a viable option.

Although tides are rarely as extreme as they are in the Bay of Fundy, they are always a force to be considered in coastal operations. Julius Caesar lost part of his fleet to a night of high tides when he was in unfamiliar waters off the coast of England. As recently as World War II, some of the disastrous landings on Pacific islands were the result of insufficient knowledge of local tides. A good tide table is indeed a good friend.

Learning Objectives

After completing the reading assignment and viewing the program, the student will be able to:

☐ Explain the principal tide-generating forces.

☐ Describe the conditions that produce spring tides and neap tides.

☐ Compare and contrast the type of tide that occurs on the west, east, and Gulf coasts of the United States.

☐ Use a tide table to predict times and heights of high and low tides.

☐ Locate regions such as the Bay of Fundy, Mont-Saint-Michel, and others having exceptional tides, and try to determine causes.

☐ Describe tidal bores, ebb and flood currents, maelströms.

☐ Correctly use the vocabulary related to tides.

☐ Recognize the importance of tides to both human activities and intertidal organisms, and evaluate tidal power as a viable source of energy.

Key Terms and Phrases

Tide The periodic rise and fall of sea level resulting primarily from the gravitational attraction of the sun and moon and the rotation of the earth. Tides are a complex phenomenon influenced by as many as 150 different tide-generating forces. Tides are important in construction of all coastal installations, in shipping, in flushing harbors and bays, and in keeping beaches clean. Intertidal organisms regulate their life activities to the ebb and flow of the tide.

Tidal range The difference in height between consecutive high and low tides. The tidal range may vary from zero at an amphidromic point to as much as 18 m in the Bay of Fundy. The range will also vary at one locality as the tides go through their monthly cycle.

Lunar day The period of one full tidal cycle: 24 hr, 50 min, 47 sec. The lunar day is slightly longer than the solar day (24 hr), which is based upon the rotation of the earth. During the time the earth is turning on its axis, the moon is advancing forward in its orbit, and the earth must turn an additional 50 min to see the moon in the same place it was the night before. The tides also lag about 50 min each day.

Spring tides The maximum range of tides that occur every two weeks during the new and full moon phases when the earth, sun, and moon are in a line. The highest high tides and lowest low tides may be expected during this part of the lunar cycle.

Neap tides The minimum range of tides that occur during either of the quarter moons when the sun and moon are at right angles to the earth. Neap tides occur during the week between the spring tides and are a period of lowest high tides and highest low tides.

Diurnal tides A tidal cycle consisting of one high tide and one low tide in each 24-hr, 50-min period or lunar day. Diurnal tides are characteristic of part of the Gulf of Mexico.

Semidiurnal tides A tidal cycle consisting of two high and two low tides of equal height per lunar day, characteristic of the East Coast of the United States.

Semidiurnal mixed tides Two high tides and two low tides that are of unequal height per lunar day; characteristic of the West Coast of the United States.

Tidal bore A steep-fronted tide crest or wave that moves up river in association with a high tide. For a strong bore to develop, the tidal range must be at least 6 m or 20 ft and the river valley must be long, funnel-shaped, and relatively shallow.

Grunion A small silvery fish (up to about 7 in. long), which spawns on the beaches of California from south of San Francisco to Magdalena Bay in Baja California. The fish come up on the beaches from March through July on the second, third, and fourth night following the new and full moon. The timing is very predictable relative to the time of high tide on any particular beach. How the grunion know exactly when to arrive is not understood.

Before Viewing

☐ Read pages 267–285 in Garrison, or pages 204–226 in Ingmanson & Wallace. Carefully study the diagrams in either text to understand tides and tide-generating forces.

☐ In the Glossary of the text, look up all words or expressions starting with *tide*. Also look up:

amphidromic point	mean lower low water
amphidromic region	mean sea level
ebb current	mean water level
flood current	neap tide
mean high water	semidiurnal tide
mean low water	spring tide

The vocabulary relating to tides is specific, and learning the words and expressions is helpful in understanding this complex but interesting subject.

☐ Watch Program 14: "The Ebb and Flow."

After Viewing

Reread the Overview and the key terms and phrases in Lesson 11 of this guide. The diagrams in the texts are very helpful and interesting. Note Figures 11.24 and 11.25 on pages 276 and 277 in Garrison, or Figure 12.19 on page 214 of Ingmanson & Wallace. Can you distinguish between the semidiurnal tides and the semidiurnal mixed tides? Check Figure 11.26 (page 278) of Garrison, or Figure 12.20 (page 215) of Ingmanson & Wallace which show tides on various coastlines. Note especially the distribution of tidal types along the coasts of the United States.

Complete the following exercises:

1. Try to explain the tides according to Newton's theory of gravitation. Explain why the moon exerts a stronger pull on the earth than does the sun.

2. Make a diagram showing the relative positions of moon, Earth, and sun during spring tides and during neap tides. Be sure to show the bulge of water on the side opposite from the moon. Try to explain this second high tide.

3. What local factors might influence the tidal range? Why are the tides so high at the Bay of Fundy? Why is the horizontal range so dramatic at Mont-Saint-Michel in France? See page 205 of Ingmanson & Wallace for excellent photographs of this phenomenon.

4. Learn to use a tide table by reading Q-and-A #3 on page 286 of Garrison, or by examining Table 12.2 on page 215 of Ingmanson & Wallace. Note that time is shown using a 24-hour clock. If you are reading Ingmanson & Wallace, notice that San Francisco has semidiurnal mixed tides and the two high tides are of different heights. A similar situation occurs at Los Angeles (page 277 of Garrison). The height of the tide is shown in meters above a certain level designated "datum." What is datum on this tide table? (Read the fine print at the bottom of the table.) *Tidal datum* is a level from which heights and depths of the sea are measured. Datum on this table is listed as "mean lower low water," which is an average of all the lower of the low tides. MLLW is datum on all West Coast tide charts. On the East Coast mean low water (MLW) is used. A comparison (in either text) of the tides of the northeast coast and the Pacific coast will explain this difference.

5. From Table 12.2 in Ingmanson & Wallace, determine when spring tides were present and when there were neap tides. The negative sign before some of the low tide figures indicates the water level was below datum. Note Saturday, December 22. What was the time interval between the two high tides? What should it be, theoretically? (Half a lunar day.) Now check the time interval between the low tides. (Now you see why a

tide table is a necessity for mariners, and how those 150 variables can affect the tides.)

6. From Table 12.2 in Ingmanson & Wallace, determine the tides on Christmas Day, December 25. What was the maximum range of the tides from highest high to lowest low? (From 1.8 m to –0.2 m equals 2.0 m, or about 6.6 ft.) What was the maximum range on Saturday, December 22? Are these spring or neap tides? Negative or minus tides are usually accompanied by highest high tides on the same day and are a clue to spring tides. If you looked at the moon that night, what are two possibilities that you might see?

7. Describe a tidal bore. Discuss the height and velocity of some famous tidal bores.

8. In an energy-hungry world, why are we so slow to develop tidal power? Discuss the requirements for a tidal power generating plant, and describe some of the problems.

Optional Activities

1. After you are familiar with how the information is presented in a tide table, check the tide tables in your daily newspaper. Are you having spring or neap tides? Also look at the phases of the moon, which are usually shown on the same page. During which phases would it be best to go clamming at the beach? Besides clam diggers, who else might be interested in using the tide tables?

2. If you live in a coastal area, go down to the docks or to a pier and ask for a tide table at the local bait or tackle shop. (These are usually free.) Take it with you to the beach and watch the coming and going of the tides during the day. Check the tide table for the lowest low tide of the month, especially in December and June, and go to the beach or the tide pools that day. A whole new world will be revealed, the seldom-seen life of the lower intertidal zone. Fascinating!

Self-Test

1. The primary force(s) that cause(s) tides in the sea is (are)
   a. coastal earthquakes and landslides
   b. wind and storms at sea
   c. the gravitational attraction of the moon and sun
   d. the gravitational attraction of Mars and Venus
   e. the rotation of the moon on its axis

2. Spring tides can be expected
   a. every spring around April or May only
   b. every weekend during the summer months
   c. one week per lunar month
   d. every other week during the lunar month
   e. every other month during the year

3. During spring tides, clam diggers and marine biologists are very active because
   a. the clams and other animals spring into the air and are easily captured
   b. the lowest zones of the beach, and intertidal areas, are accessible
   c. the height of high and low tide is almost the same, with little wave action
   d. the intertidal zone remains underwater, and they can watch the animals feed
   e. all of the above

4. The tide patterns characteristic of the Pacific coast are
   a. semidiurnal mixed tides
   b. diurnal tides
   c. semidiurnal tides
   d. reversing tides
   e. amphidromic tides

5. The tides at any one locality will result from the interaction of
   a. sun, moon, and earth
   b. elliptical orbits of the moon and earth
   c. rotation of the earth
   d. shape, size, and depth of the basin
   e. all of these

6. A foaming, churning wall of water moving into a bay or up a river is a
   a. spring tide
   b. mixed tide
   c. tidal wave
   d. tidal bore
   e. tidal range

7. During one lunar day, the harbors along the East Coast of the United States
   a. will experience one high and one low tide
   b. one high and two low tides
   c. two high tides and two low tides of equal height
   d. two high tides and two low tides all of different heights
   e. two high tides and three low tides of equal height

8. A very small tidal range is characteristic of
   a. Hawaii
   b. islands near the center of the ocean basins
   c. the Mediterranean
   d. the Gulf of Mexico
   e. all of the above

9. If the moon is full and in perigee, and the earth is in perihelion, the tides are most likely to be
   a. extreme spring tides
   b. lower-than-usual spring tides
   c. minimal in range
   d. higher-than-usual neap tides
   e. completely suppressed for one day

10. Grunion are rather famous, considering their small size and nondescript appearance. They are best known for their
    a. reputation as a gourmet delicacy
    b. sharp teeth and voracious appetites
    c. habit of cleaning other fish
    d. unusual breeding behavior that is tied to tidal cycles
    e. ability to breed in rivers and streams

Supplemental Reading

Greenberg, David A., "Modeling Tidal Power." *Scientific American*, vol. 257, no. 5 (1987), 128–131.

Lesson Fifteen   Plankton: Floaters and Drifters

Overview

The community of drifting organisms called plankton is as inconspicuous in the ocean as it is important. *Plankton* is the term applied to animals and plants that live suspended in water. Although these plants and animals may be moved over considerable distances in a day's time, they are unable to make any effective movement against the currents. Plankton is distinguished from nekton, which is composed of strongly *swimming* animals. Incidentally, size is not a factor in determining whether an organism is classified as plankton. Drifting organisms come in all sizes, though most tend to be small.

There are two major divisions of planktonic organisms: phytoplankton and zooplankton. The phytoplankton are the plants of the drifting community, and the zooplankton are the animals. These names are derived from the Greek *phytos*, which means "plant," and *zoion*, which means "animal." Very small phytoplanktonic or zooplanktonic organisms are called nanoplankton, a word derived from the Latin *nanus*, meaning "dwarf."

Two phytoplankters are especially important: diatoms and dinoflagellates.

Diatoms are the earth's most successful and abundant plants. They are responsible for most of the food produced from sunlight and inorganic nutrients in the seas, and for most of the oxygen production. Because of their large influence, they may be considered the earth's most important form of plant life. Diatoms evolved in Cretaceous times, relatively recently in the history of the earth. The appearance of so efficient a primary producer created major changes because of the oxygen added to the atmosphere and the larger amount of available food upon which food chains might be based.

These beautiful single-celled plants have a transparent, two-part, glass shell that fits together like a pillbox to contain the living protoplasm. The shell is perforated, usually in regular patterns, to permit the flow of materials between plant and ocean. The largest diatoms are just large enough to be visible to the unaided eye when viewed under a strong light. Figures 14.10 on page 358 in Garrison, or Figure 14.4 on page 273 of Ingmanson & Wallace, show typical diatoms.

Dinoflagellates, as a group, are as ancient as the diatoms are modern. The dinoflagellates probably were among the first oceanic plants to evolve about 2 billion years ago. Instead of a symmetrical pair of glass shells, the dinoflagellates are covered with either a proteinlike or cellulose compound, and sometimes both. The dinoflagellates are named for the presence of two whiplike flagella with which they can propel themselves for limited distances through the water. Under favorable conditions, dinoflagellate populations may reach 6,000,000 organisms per liter (1 L = 1.057 qt)! When present in such large numbers, some of these organisms form a "red tide" that may be toxic to other forms of marine life. Dinoflagellates also produce some of the luminescence in the oceans. Dinoflagellates are not as efficient in food production as are diatoms, and, on the average, are smaller. Figure 14.12 on page 360 of Garrison, or Figure 14.6 on page 274 of Ingmanson & Wallace, show typical dinoflagellates.

These two major groups of phytoplankton, along with some other planktonic plants, are responsible for about 95 percent of the primary productivity of food molecules occurring in the ocean. Attached benthic plants (seaweeds) account for the other 5 percent. Although many factors must be present for any plant to manufacture food through the process of photosynthesis, the two most important factors are the presence of sunlight and the availability of the appropriate nutrients. Because of their dependence on solar illumination, the phytoplankton live only in the uppermost sunlit layer of the ocean. When nutrients in the upper layer become depleted by their activity, the plants become less active and passively await a change in conditions as they drift with the currents. Reproduction may occur at this time. An upwelling of nutrient-rich bottom water would be especially welcome at such a period in the life cycle.

The production of food molecules by photosynthesis occurs primarily in those areas where the water is warm and nutrients are plentiful. Unfortunately, very few such idyllic spots exist. The tropics have warm water and an abundance of sunlight but usually lack the requisite nutrients. The polar regions have many nutrients but lack heat and sun. The greatest productivity, therefore, occurs in the nearshore temperate zones where vertical water movement can bring nutrients to the surface to combine with adequate sun and water temperature. Not coincidentally these areas are rich fishing grounds teeming with oceanic birds. Phytoplankton forms the base of virtually all oceanic food chains.

The zooplankton, the primary consumers of the ocean, graze on the phytoplankton much as a cow grazes on grass. The variety of zooplankton is astonishing. Every major animal phylum is represented here. Size is no guarantee that an organism can outswim the currents, so even large medusae (jellyfish) are classed as zooplankton. These may have a bell diameter in excess of 6 ft and trail 50-ft tentacles. Most zooplankton representatives are barely visible to the naked eye, but are about ten times larger on the average than the phytoplankton on which they graze.

The zooplankton is divided into two groups. Temporary members of the community, called meroplankton, are usually immature animals that will either grow out of the plankton category when they become strong swimmers, or settle out to become bottom dwellers or fouling organisms. Permanent zooplankton are called holoplankton. Although meroplankton is abundant in coastal regions, the holoplankton is by far the larger group.

The most abundant zooplankton group is the class Crustacea of the phylum Arthropoda, occupying a niche in the ocean equivalent to insects on land. They are found in both the meroplanktonic and holoplanktonic categories. Familiar nonplanktonic representatives of this class of organisms include lobster, shrimp, and crab. Certain planktonic forms somewhat resemble shrimp in appearance. Curiously, the largest animals ever to live on the earth, the great blue whales, eat these small planktonic crustacea.

Direct consumption of small plankton by large animals, though, is comparatively rare. The whale is a monumental exception to the usual course of marine food chains. Food chains are usually longer and more complex in nature, as you will learn in Lesson 20.

Plankton is the most important lifestyle group in the ocean, forming the productivity base for nearly all marine food chains. There is considerable potential for the direct harvesting of plankton for human consumption, a promising possibility for the undernourished populations of the world. How such large-scale harvesting would affect the already stressed ecological balances in the seas is not well understood. We have much to learn about this beautiful, important, and complex community of drifters.

Learning Objectives

After completing the reading assignment and viewing the program, the student will be able to:

☐ Contrast the planktonic lifestyle with the nektonic lifestyle.

☐ Compare phytoplankton with zooplankton.

☐ Describe the basic characteristics of diatoms and dinoflagellates.

☐ Informally define primary productivity, name some factors that influence it, and describe the relative contributions of phytoplankton and fixed plants (that is, seaweeds).

☐ Contrast meroplankton with holoplankton.

☐ Indicate the places on the earth, in general terms, where plankton is most successful and abundant, and list the three most important factors that influence the growth and success of phytoplankton.

☐ Understand why, contrary to popular belief, oceanic productivity in the tropics is usually very low.

Key Terms and Phrases

Plankton Floating or drifting organisms in the water whose movements from place to place are due to the mass movements of the water rather than to their own swimming efforts.

Phytoplankton The plant forms of plankton. They are the basic synthesizers of organic matter (by photosynthesis) in the open ocean. The most abundant and important phytoplankton are the diatoms.

Zooplankton The animal forms of plankton. They include crustaceans such as copepods and euphausiids, jellyfish, certain protozoans, worms, mollusks, and the eggs and larvae of many benthic and nektonic animals.

Primary productivity The amount of organic material synthesized by organisms from inorganic substances. This nearly always is accomplished by photosynthesis.

Photosynthesis The manufacture of carbohydrate food from carbon dioxide and water in the presence of chlorophyll by using light energy. Oxygen is released as a waste product in these reactions. Photosynthesis is the means by which solar energy is made available to power virtually all of the biological systems of the earth, on both land and sea.

Diatom Earth's most successful and abundant plant. It is one of a class of single-celled microscopic phytoplanktonic organisms possessing a wall of overlapping halves impregnated with silica (glass).

Dinoflagellates One of a class of single-celled microscopic organisms grouped with the phytoplankton. These organisms are occasionally responsible for the formation of red tides.

Plankton bloom An enormous concentration of phytoplankton in an area, which is caused by either an explosive or a gradual multiplication of organisms (sometimes of a single species). Plankton blooms usually produce an obvious change in the physical appearance of the sea surface, such as discoloration. Blooms consisting of millions of cells per liter (1 L = 1.057 qt) have often been reported.

Meroplankton Members of the plankton community that spend only part of their life cycle as plankton (example: larval fishes).

Holoplankton Members of the plankton community that spend their entire life cycle as plankton (example: copepods).

Copepoda The most abundant small zooplanktonic arthropod. Copepods are the most frequently encountered "primary consumers" of the seas, and graze upon diatoms and dinoflagellates. (Note: Phylum Arthropoda includes such animals as shrimp, lobster, crab, and insects.) Copepods form the second link in many oceanic food chains and food webs.

Before Viewing

☐ Read the sections in Ingmanson & Wallace that discuss plankton:

— pages 268–269, "Classification and Terminology"

— pages 272–277, "Single-Celled Organisms: The Kingdom Protista"

— pages 293–298, "Crustaceans"

— pages 325–328, concerning the feeding behavior of cetaceans

— pages 334–337, "Density and Viscosity"

— pages 340–350, "Plant Productivity and Food Chain Dynamics"

— pages 352–353, Figure 16.14

Or read the sections in Garrison that discuss plankton:

— pages 355–367, "Plankton;" Figure 14.18 on page 366 has a drawing of one of the most important zooplanktonic forms, genus *Calanus*

— page 403–408 on the relationship between great whales and zooplankton

☐ Watch Program 15: "Plankton: Floaters and Drifters."

After Viewing

Reread the text sections, being sure you understand the key terms and phrases, and complete the following exercises:

1. What factors, in order of importance, affect the growth of phytoplankton?

2. Why is plankton so successful? What is there about the planktonic lifestyle that lends itself to great biological success?

3. Where would you guess the greatest amount of photosynthesis occurs over a year's time – in the oceans of the world or on the land?

4. Why do you suppose the great whales have evolved a feeding method that permits them to eat from so low a position in the food chain/web? In other words, why do such big animals eat such small food?

5. Diatoms are very successful, yet they have shells made of heavy glass. If they are photosynthetic, they need to stay near the surface. Does not the heavy glass weight them down and send them toward the bottom? If so, how do they counter this difficulty if they can't swim?

6. Can phytoplankton ever get too much light? Too little light?

7. What is the difference between a "food chain" and a "food web"?

Answer Key for Exercises

1. Phytoplankton require three things to keep them going. The first of these is sunlight. Without sunlight a photosynthetic organism would eventually die from the lack of ability to synthesize food. (For additional information on this see the answer to question 6.) A "co-first" in importance with sunlight is the presence of inorganic nutrients. These fertilizers are critically important in allowing the organism to generate proteins and other compounds it requires for life. Next in importance is water temperature. In general, the higher the water temperature the greater the biological activity. Therefore, higher water temperatures lead to higher rates of growth. However, spectacular blooms can occur at lower water temperatures as long as sunlight and nutrients are available. This will be discussed in greater detail in Lesson 24, "The Polar Seas."

2. Plankton is successful because of its size. Microscopic plants don't require the elabo-

rate pumping and support systems of large terrestrial plants. Their small size allows easy diffusion of required nutrients into the single cell and prompt transport of wastes out. With a glass shell (transparent to sunlight) to protect them, diatoms are especially successful. The recent evolution of diatoms was a high point in the development of oceanic plant life, and the additional oxygen made available to the atmosphere by these plants may have influenced the success of life on land. Of course, more plants may have led to more animals, and after a few hundred million years of evolution the planktonic lifestyle is now elegantly tuned and interlocked into the complex and productive system we now observe in the world ocean.

3. Estimates vary, but the ocean is probably responsible for the majority of primary productivity on the earth. It was once thought that the ocean was responsible for many times the productivity of the combined land areas. However, recent research has indicated that previous estimates of land productivity were probably underestimated. The great rain forests of the earth contribute more productivity to the total planetary figure than was previously supposed. Even so, the ocean in its totality probably generates more food and oxygen from inorganic material (using photosynthetic processes) than does the land.

4. The lower on the food chain an organism eats, the more energy initially bound by photosynthesis is available to it. Lots of energy is lost in complex food webs because lots of organisms must be supported. Whales get around this difficulty by eating from a fairly low point in the chain, therefore obtaining bound energy with more efficiency.

5. Diatoms have a trick. They sometimes use a small droplet of oil to offset the heavy shell. Oil floats, glass sinks, and the correct amount of both will result in neutral buoyancy. In fact, the diatom can precisely adjust its buoyancy so as to be in exactly the correct depth to receive the right amount of sunlight for the prevailing conditions.

6. Yes, the amount of sunlight can often be too great or too little. If the phytoplankters are

too near the surface of the ocean at noon, the excess sunlight can actually stop photosynthesis. If they are too deep, there may be adequate illumination for only an hour or so per day. The trick is to be at a depth adequate to protect from overexposure to bright light, but still allow photosynthetic activity to occur over a large number of hours per day. This depth is somewhere around 10 m (30 ft) for many diatoms in most temperate oceans, though it varies greatly. Many dinoflagellates can move vertically in the water to optimize the amount of sunlight received. They are still plankton, of course, because although they can move vertically, they have no ability to outswim the currents with which they are drifting.

7. Semantics, mainly, but the image of a chain is linear. The implication of a food chain is that one organism eats only one kind of organism and is eaten only by the next specific kind of creature, and so on, up the single line of the chain. A more accurate representation of what happens in the marine ecosystem is represented by the food web, an intricate mesh of food chains – a construct with crisscrossing opportunities for eating whatever tasty morsel blunders along.

Optional Activities

1. If you live near any large body of water (fresh water or ocean water), you will probably be able to find some plankton. Plankton is easily captured with a plankton net. Lacking a professional net, you can make a perfectly adequate substitute from the toe end of a woman's nylon stocking. Drag this through the water for a few minutes and empty the contents into a jar along with some water from the same area. Observe your catch with a strong light, such as that from a slide projector. You will be able to see many zooplankters and some larger phytoplankters in the beam. If you have a magnifying glass or access to a microscope, so much the better.

2. Some museums have glass models of plankton. You might enjoy the lovely shapes and delicate construction of these glass blowers' nightmares. They are often excellent and accurate reproductions of the real organisms.

3. The Mall, in Washington, D.C., is a most unusual location for one of the most remarkable demonstrations of planktonic primary productivity in the United States. The Smithsonian Institution's National Museum of Natural History has on display the only living, functioning, and balanced coral reef outside the tropics. To make the reef function properly, plankton must be cultured and made to thrive in the enclosure. This is done by marine biologists, students, and volunteers on the staff who take great care in precisely reconstructing the physical and chemical requirements for plankton productivity. Lights of exactly the right color turn on and off to replicate the gradual increase of light at dawn (and decrease at dusk), and nutrients are carefully added in exactly the right proportion to ensure growth. Waves are even generated to aerate the water and make the fish, coral organisms, and other specimens feel right at home in Washington.

Literally and appropriately, right around the museum corner from the reef is another kind of plankton spectacular: a full-size model of a blue whale, an organism that depends on zooplankton for its food. Don't miss an opportunity to see these exhibits if you visit the nation's capital.

Self-Test

(More than one answer may be correct.)

1. Small plants that drift with ocean currents are called
   a. plankton
   b. phytoplankton
   c. zooplankton
   d. meroplankton
   e. holoplankton

2. Small drifting larval forms of animals that will soon "settle out" to become benthic organisms are known as
   a. plankton
   b. phytoplankton
   c. zooplankton
   d. meroplankton
   e. holoplankton

3. Large jellyfish are categorized as
   a. plankton
   b. phytoplankton
   c. zooplankton
   d. meroplankton
   e. holoplankton

4. Diatoms are included in which of the following groups?
   a. plankton
   b. phytoplankton
   c. zooplankton
   d. meroplankton
   e. holoplankton

5. Which of these factors would probably *not* significantly affect the primary productivity of an ocean area?
   a. a heavily overcast day after a week of bright sunshine
   b. sudden lack of inorganic nutrients
   c. an abrupt drop in water temperature
   d. the presence of a pod of baleen whales feeding in the area
   e. the introduction of large amounts of chemical detergent from a chemical-carrying ship breaking up in the area

6. Earth's most successful and abundant plants are
   a. diatoms
   b. dinoflagellates
   c. seaweeds
   d. rice (terrestrial/aquatic)
   e. marine grasses

7. Seaweeds represent about _____ percent of the total oceanic productivity.
   a. less than 1
   b. 5
   c. 25
   d. 50
   e. nearly 100

8. Of the following which is the earth's most primitive life form?
   a. diatoms
   b. seaweeds
   c. dinoflagellates
   d. marine grasses
   e. none of the above

9. The most abundant zooplankton group is within
   a. phylum Mollusca
   b. phylum Chordata
   c. phylum Coelenterate (Cnidaria)
   d. phylum Echinodermata
   e. phylum Arthropoda

10. What is meant by "compensation depth"?
    a. the depth at which plankton can live
    b. the depth at which phytoplankton can live
    c. the depth at which zooplankton exactly "break even" in their energy requirements
    d. the depth at which phytoplankton exactly "break even" in their energy requirements, balancing photosynthetic production of carbohydrates with respiratory use of carbohydrates
    e. none of the above

11. Red tide is caused by
    a. diatoms
    b. zooplankton
    c. dinoflagellates
    d. oil spills
    e. copepods

Supplemental Reading

Isaacs, J. D., "The Nature of Oceanic Life." *Scientific American* (1969).

Newell, G. E., and R. C. Newell, *Marine Plankton*. London: Hutchinson, 1977.

Smith, D. L., *A Guide to Marine Coastal Plankton and Marine Invertebrate Larvae*. Dubuque, Iowa: Kendall/Hunt, 1977.

Wickstead, J. H., *Marine Zooplankton*. London: Edward Arnold, 1976.

Lesson Sixteen Nekton: Swimmers

Overview

Nektonic organisms are pelagic organisms that are swimmers. The term *nekton* contrasts nicely with *plankton*, the term used to describe drifting organisms that are at the mercy of the currents.

Nektonic animals (swimmers) are adapted for self-propelled movement through the water. Their shape is generally streamlined, and a significant percentage of their body weight is muscle. Their particular form depends to a large extent on what they eat. In fact, it is often easy to deduce a swimming organism's diet from its shape and gross anatomy.

Any swimming organism must maintain a specific level in the water if it is to compete successfully. Several tricks are used in accomplishing this feat, including the use of gas trapped within a special swim bladder to maintain buoyancy, the use of body construction materials with a specific gravity close to that of seawater, and active swimming to maintain a specific level.

There are five principal groups of nektonic organisms: cephalopod mollusks; fish; nektonic arthropods such as shrimp, and certain crabs such as the Galatheid arthropod *Pleuroncodes*; marine reptiles such as turtles, crocodiles, and sea snakes; and marine mammals. There are other relatively minor groups such as urchordates, pteropod mollusks (a sort of swimming snail), and more that verge on being classified as planktonic. Like the term *plankton*, *nekton* is an artificial classification, a grouping by lifestyle for convenience. These organisms do not constitute a taxonomic class, nor are they always directly related to one another by ancestry. See Item #1 in the "After Viewing" section of this lesson.

Cephalopods are an ancient subdivision of the phylum Mollusca and include such swimmers as squid and nautilus. This group represents a high point in invertebrate evolution and contains the most intelligent of the invertebrates as well as those with the most advanced eyes. Some species grow to astonishing size; for instance, the giant squid *Architeuthis* has been reported at lengths of over 15 m (50 ft).

Squid are the only small organisms ever to provide any real competition to the fish in the pelagic habitat. They move with great speed and agility by forcing water from special vents in their mantle cavities. Small fish fall victim to squid because of this speed, their efficient tentacles, intelligence, and excellent eyesight.

Fish themselves are the most abundant and successful of the vertebrate organisms. Over 30,000 species of fishes are now known. Generally they are categorized in three major groups within the subphylum Vertebrata, phylum Chordata. Starting with primitive animals and moving toward the more advanced in the evolutionary scale, the groups are arranged as follows:

☐ class Agnatha (Cyclostomata): hagfishes and lampreys

☐ class Chondrichthyes: sharks, skates, and rays

☐ class Osteichthyes: modern fishes with bony skeletons such as tuna, salmon, and goldfish

Hagfish and lampreys are strange primitive fishes that lack jaws and have skeletons of cartilage rather than bone. Some lampreys attach themselves to the side of another fish and feed on the living flesh of the host. They entered the Great Lakes when canals opened the lakes to the St. Lawrence Seaway, and as a result the fresh water fish in the lakes, a valuable resource, were decimated. The hagfish demonstrates an interesting defense mechanism that predators find uninviting. It consists of enveloping itself in a thick coating of slippery mucus. To free itself of the mass and clear the gill openings, it occasionally ties itself in a loose knot passing the knot down its snake-shaped body. A fresh mucal shield is then generated.

The sharks, skates, and rays are fish that have jaws but also have a cartilaginous skeleton. In addition the skates and rays have a flattened body form and are well adapted to living on the ocean bottom.

Sharks, on the other hand, are active swimmers that range the seas from the shallow surf to mid-water depths of the open sea. They are an amazingly varied group of fishes that range in size from small sand sharks that feed in the subtidal zone to the giants of the seas, the whale shark and the basking shark. The gentle basking shark, which may be 17 m long (56 ft), is fortunately a plankton feeder.

Some sharks lay eggs in beautiful and unusual egg cases, and others give birth to live young. Most have the familiar streamlined body and pointed head. The hammerhead shark has a bizarre cross-bar or "hammer" across the head with the eyes on each end. The exact function of this unusual feature is not known, but it may provide excellent binocular vision in clear water or even a sort of "stereo smell" to sense different amounts of interesting substances in the nearby water.

Shark attacks on humans have been widely publicized in films, television, and popular press. Actually, they are relatively infrequent on a worldwide basis. More than 80 percent of shark species are less than 2 m (6.6 ft) long as adults, and only a minority of the remaining 20 percent are aggressive toward people. Like other cartilage fish, they are not intelligent and certainly do not hold grudges or otherwise behave in the malignant ways vividly fictionalized in recent popular movies. Still, some sharks are indeed dangerous, the legendary white sharks of genus *Carcharodon* perhaps most of all (see pages 393–394 of Garrison, or Figure 15.4 on page 317 of Ingmanson & Wallace). These divers' nightmares routinely attain lengths of 7 m (23 ft) and weigh 1,400 kg (2,800 lb). They are not white, but a grayish brown or blue above a creamy lower half. Another dangerous relative, the Mako shark, reaches 4 m (13 ft) and is known to attack small boats as well as people. These and other predatory sharks are attracted to vibrations or body fluids in the water, which

they detect with sensitive organs arrayed in lines beneath the surface of their skin. Their customary prey is fish and marine mammals.

On the other hand, human attack on sharks is becoming increasingly frequent, as shark meat is added to more menus. Thresher sharks, in particular, are noted for fine flavor and texture. Shark-fin soup, of course, is an ancient delicacy.

Fish with bony skeletons (Osteichthyes) are the most familiar vertebrate animals of the sea. They appear far back in the fossil record and over geologic time have radiated to fill many ecologic niches. Today fishes live at all depths, from the tidepools populated with little blennies, gobbies, and clingfish, to the greatest depths, the home of strange black, soft-bodied, luminescent forms.

The variety of body shape, color, and features in modern fishes seems endless, but certain features are characteristic of the group as a whole. Most fishes swim through the water by moving their whole body, not just by wiggling their fins. The fastest swimmers belong to the mackerel-tuna family and to the family of sailfish. Some members of the sailfish group have been clocked at an astounding 96 km/hr (60 mph). Tuna of about 1 m long have been clocked on short runs at 75 km/hr (45 mph). The barracuda can speed along at about 40 km/hr.

Although fishes are generally considered "cold-blooded" organisms, it has been discovered that some members of the mackerel-tuna family and a few species of sharks have internal temperatures above that of the surrounding water. The internal body temperature of the tuna is about 32°C and remains relatively stable even in cold water (normal human temperature is 37°C). This high internal body temperature permits faster and more efficient food-to-musclepower energy conversion, and allows these predatory fish to exist successfully in a niche similar to that occupied by the smaller marine mammals.

The color of swimming animals can be striking. There are many very practical reasons for this, including the need to blend in with surroundings (cryptic coloration), attempts to camouflage the body by means of stripes and spots that break up the shape lines of the body (disruptive coloration), species recognition for schooling and mating (identification coloration), countershading for less conspicuous movement in the open water (obliterative coloration), and the color-shape patterning associated with mimicry. Among swimmers it is the fish that exhibit the most elaborate coloration. The squid and marine mammals rely in part on intelligence and maneuverability to help avoid predation.

Smaller fishes often form groups known as schools for mating purposes, for food gathering, and for protection. Predators will sometimes mistake 1,000 small fish for one large organism and retreat. As fish grow larger they tend to school less.

Many swimming arthropods inhabit the oceans, but most of these are planktonic. Of the swimming arthropods you are probably most familiar with the larger nektonic shrimp, which live in abundance in the waters of the Gulf of Mexico and in other productive temperate and tropical locales. Some marine nektonic reptiles will be discussed in Lesson 17. Marine mammals will be covered in Lessons 18 and 19.

Learning Objectives

After completing the reading assignment and viewing the program, the student will be able to:

☐ Distinguish between plankton and nekton; describe the nektonic lifestyle in general terms.

☐ List the five principal categories of swimming organisms.

☐ Cite some mechanisms by which swimming organisms maintain their position in the water column.

Identify the three taxonomic categories of fishes and give examples.

Key Terms and Phrases

Pelagic Referring to open-ocean water. The opposite of pelagic is *neritic*, referring to water near the shore. Another contrasting word is *benthic*, which means on (or in) the bottom.

Nekton Pelagic animals that are active swimmers, such as most of the adult squid, fish, and marine mammals.

Plankton The passively drifting or weakly swimming organisms in marine and fresh waters. Members of this group range in size from microscopic plants to jellyfish measuring 2 m

(6.6 ft) across. "Plankton" is not a taxonomic category, it is a lifestyle.

Cephalopod(a) One of the five major divisions of the phylum Mollusca, which includes squid, octopus, and nautilus.

Vertebrate/invertebrate Vertebrate organisms have backbones as adults; invertebrates do not have backbones.

Cold-blooded (Ectothermic) Organisms whose internal temperature is controlled primarily by the temperature of the environment. Warm-blooded organisms, such as humans, have a stable internal temperature, which is, under normal conditions, independent of external temperature.

Diversity Variability in form within a group.

Taxonomic Refers to the arrangement law used by biologists to categorize organisms within groups. The categorization is done in part on the basis of evolution from common ancestry.

Arthropod A member of the phylum Arthropoda, the world's most successful group of animals. Arthropods include lobsters, shrimp, crabs, all insects, most zooplankton, perhaps 1,800,000 species in all.

Before Viewing

☐ Read Chapter 15, pages 376–400 in Garrison, or pages 306–316 of Chapter 15 in Ingmanson & Wallace. Some nektonic organisms are also covered in Ingmanson & Wallace's Chapter 14. Glance through the appropriate chapter to see what swimmer you can discover.

☐ Watch Program 16: "Nekton: Swimmers."

After Viewing

Reread the text section, being sure you understand the key terms and phrases, and complete the following exercises:

1. *Nekton* is a bothersome term to many marine biologists. Why do you suppose this is so?

2. Tropical fish are noted for their brilliant coloration. Why do you think tropical fish

should be so vivid and their temperate counterparts rather more drab?

3. Squid are the only small animals in the sea that have threatened the ascendancy of fish. If squid have such relatively sophisticated brains and such good eyesight, and have been around on the earth for so long a time, why are they not more successful? There are some 30,000 species of fish, but only about 600 species of squid.

Answer Key for Exercises

1. The term *nekton* troubles many marine biologists because so many organisms are indiscriminately lumped into the category. There are many organisms that are nektonic for part of their life and benthic, or even planktonic, for another part. This category, like the term *plankton*, is only a description of a lifestyle – and the lifestyle may only be temporary in some cases.

2. Fish are not the only tropical organisms that are brilliantly colored. One sees flashes of color from tropical birds, marine invertebrates, and other kinds of animals. Brilliant coloration seems a hallmark of the tropics. The reason for this vivid display lies in the tremendously large number of species of animals that live in the tropics. With so many species in the area, color and shape displays are used to attract members of the same species for reproductive activity. In a sense, the brilliant coloration is a sort of specific advertisement that a specific animal is a member of a specific species. Members of similar species do not respond to the exact pattern or display behavior of a different species.

3. The main advantage fish have over the cephalopod mollusks (squid, nautilus) is one of architecture. Fish are members of the advanced vertebrate class within the phylum Chordata; squid are members of the ancient phylum Mollusca. The cephalopods are analogous to an old-design automobile chassis on which designers have hung all sorts of elaborate and "new-fangled" options; the chordates are analogous to a new automobile blueprint that is derived from a clean sheet of paper and takes full advantage of modern technology. The musculature of the two organisms is very different; the fish is a stronger and generally more agile predator. The reproduction of fishes is more efficient, and the neural arrangements are more sophisticated. The circulatory system and respiratory arrangements are also more advanced in fish.

Squid do remarkably well with the primitive bodies they have perfected, but evidently they do not do well enough. The fact that squid are losing the battle of the oceans to the fish is obvious in the fossil record. In the Paleozoic and Mesozoic eras there were thousands of species of squid, but today only a small sample remains. The distant future of the squid does not look particularly good.

Optional Activities

1. Visit a large aquarium or oceanarium and watch the swimming behavior of fishes. Notice that the whole body of most fishes undulates, not just the fins. Note the streamlining detail.

2. Investigate the internal anatomy of a squid or octopus. This can be done by looking in a good zoology textbook or by preparing a commercial squid for cooking. Squid are available at many fish markets or bait stores.

3. When you next see a living fish (or a photograph of a fish), try to explain how its color patterns would be of benefit.

Self-Test

1. The term *nekton* refers to
   a. organisms that drift with the currents
   b. organisms that live on or in the bottom
   c. plants only
   d. organisms that swim actively
   e. fishes and cephalopods

2. Which of these is (are) generally characteristic of nektonic organisms?
   a. high muscle-to-body-weight ratio
   b. utilization of some mechanism to adjust position vertically in the water column
   c. speed and agility
   d. relatively high intelligence within taxonomic groups
   e. all of the above

3. The most abundant nektonic organisms are
   a. fish
   b. marine mammals
   c. cephalopods
   d. sea cucumbers
   e. shrimp

4. Among the most highly evolved marine invertebrates are
   a. sea anemones
   b. tuna
   c. small whales
   d. sea cucumbers
   e. squid

5. Coloration in fish is used for
   a. species identification for mating or schooling
   b. blending into the surroundings
   c. disrupting the shape and outline of the fish to fool potential predators
   d. assisting in feeding by permitting the predator to resemble another organism or object
   e. all of the above

6. The most primitive group of living fishes is the
   a. class Agnatha
   b. class Chondrichthyes
   c. class Osteichthyes

7. Most sharks are dangerous to humans.
   a. true
   b. false

8. Which of these nektonic organisms has the highest internal body temperature?
   a. a tuna
   b. a squid
   c. a small whale
   d. a shrimp
   e. a lamprey

9. The most advanced and highly evolved nektonic organisms are in which group?
   a. class Osteichthyes, phylum Chordata
   b. whales
   c. class Cephalopoda, phylum Mollusca
   d. sea cucumbers
   e. phylum Arthropoda

10. Are all nektonic organisms predators?
    a. yes
    b. no

11. Which of the following statements is true of squid?
    a. Squid are a highly successful nektonic group.
    b. They are among the most sophisticated animals in the oceans.
    c. They are an ancient group of organisms that has found itself in sort of an evolutionary "dead end."
    d. They are equal to fish in nearly all ways.
    e. Squid do not have good eyesight at close range.

Supplemental Reading

Garrison, T., "Rays." *California Diver*, vol. 2, no. 1 (1986), 17–20.

Gilbert, P. W., et al., "Sharks." *Oceanus*, vol. 24, no. 4 (1982). (Overview written for the general reader.)

Gray, J., "How Fishes Swim." *Scientific American* (1957), 48–54.

Partridge, B. L., "The Structure and Function of Fish Schools." *Scientific American*, vol. 246, no. 6 (1982), 114.

Shaw, E., "The Schooling of Fishes." *Scientific American* (1982).

Stanley, S. M., "Mass Extinctions in the Ocean." *Scientific American*, vol. 250, no. 6 (1984).

Ward, P., et al., "The Buoyancy of the Chambered Nautilus." *Scientific American*, vol. 243, no. 4 (1980), 190.

Webb, P., "Form and Function in Fish Swimming." *Scientific American*, vol. 251, no. 1 (1984).

Lesson Seventeen   Reptiles and Birds

Overview

The Age of Reptiles reached the height of its success during the long Mesozoic era, 150 million years ago. The reptiles found today are remnants of what was then a much larger and more diverse population. At one time there were fish-shaped ichthyosaurs and long, "sea serpent" plesiosaurs roving the seas, but today there are only three families of marine reptiles left in the world. These are the family of Cheloniidae (sea turtles), the family Hydrophidae (sea snakes), and the family of sea lizards whose only contemporary representative is the Galápagos iguana, genus *Amblyrhynchus*.

The seven widely distributed tropical and subtropical species of sea turtles all have large limbs and nonretractable heads. They are excellent swimmers with front limbs flattened to act as oars, and hind limbs that work as rudders. These animals grow to considerable size. The Atlantic leatherback, the largest living species, sometimes grows to over 680 kg (1,500 lb) and 3.5 m (11.5 ft) in length.

Sea turtles migrate regularly between their feeding areas and the beaches on which they spawn. Some species apparently use celestial navigation, smell, and vision to zero in on a beach site close to the point where they themselves were hatched.

Sea turtles today are under considerable danger of extinction. Whalers and traders, in the days of sail, captured large numbers of turtles and stored the long-lived animals on deck as living sources of meat. The greenish fat beneath the turtle's shell has for years made them a target of poachers who sell the substance for turtle soup. The eggs are used by animals, including the human species, for food. Even though all of these activities are illegal in most of the areas frequented by sea turtles, the population still continues to shrink. Turtles are now being successfully farmed in an attempt to replenish the depleted natural population.

Probably the most recently evolved group of marine reptiles are the 50 species of sea snakes. The toxins of sea snake venom are among the most active biological poisons known, with a physiological effect similar to that of the cobra's venom. Sea snakes live primarily in the warm oceans of the Indo-Pacific, but have been reported in most of the tropical oceans of the world. These reptiles are truly marine and, with the exception of one species that lays eggs, never come to land. Their bodies are flattened for swimming, they give birth to live young, and they capture small fish in their jaws. Their small teeth hold the fish until the venom can seep into the wounds and subdue the fish. Fortunately for man, these snakes are not aggressive and lack a venom injection apparatus. However, as the fishermen of Southeast Asia know, even a tiny scratch from a sea snake tooth, which admits a tiny quantity of saliva into the wound, can be very dangerous.

Marine lizards (marine iguanas) are a spectacular reminder of the Age of Reptiles. Seeing these primitive-appearing creatures on the ashy volcanic wasteland of the Galápagos gives one a feeling of what this era must have been like. Sea lizards are large, heavy, and sluggish, extending about 1 m (3.3 ft) in length and weighing about 7–9 kg (15–20 lb). They are usually sooty gray to brown in color and lumpy, without much grace on land. In the sea, however, they move efficiently by undulatory movements of body and tail. They eat encrusting seaweeds with flattened, spadelike teeth, and they have special glands in their head that extract excess salt from their blood, allowing it to be ejected to the outside through their nostrils. They live in large colonies at the water's edge and return to these settlements after foraging a short distance off-shore for the encrusting seaweeds that form the substance of their diet.

One additional family of reptiles that is primarily estuarine, but which is occasionally found at sea, is the crocodile. Some ocean-going crocodiles have been discovered that are greater than 7 m (23 ft) in length! These monsters are found from India to northern Australia and (fortunately) are relatively rare. These crocodiles have an unusual habit: They travel in groups for fairly long distances between islands, arriving on a shoreline by surprise to prey upon the island's mammalian inhabitants. These crocodile colonies seem to track a sort of "circuit" around sets of islands, returning to the same island after many years.

At first glance, it may not seem logical to study birds and reptiles in the same lesson, but closer observation reveals many similarities. The feet and lower legs of a bird are covered with scales and terminate in claws. The reproductive physiology is basically reptilian (feathers are thought to be derived from scales), and there are many developmental and structural similarities.

Fossil records indicate that birds split off the main reptilian line some 150 million years ago with the intermediate form called *Archaeopteryx* (the word means "early flyer"). This primitive bird was probably more of a gliding feathered reptile than a true bird. In a sense, birds are specialized and perfected reptiles. Many scientists feel that modern birds represent a current end point of reptilian evolution and that they are heirs to the legendary dinosaurs.

In the ocean environment birds are a good indication of high surface productivity. Large numbers of birds are often found where large amounts of plankton support a rich surface nektonic community. Periodically they return to shore to form tremendous resting and breeding grounds. These land areas are important sources of nitrates and fertilizers and are occasionally mined.

Birds are a beautiful, spectacularly diverse and highly developed group, well adapted to their environment. Birds in the oceanic habitat, for example, do not drink fresh water. Instead they have salt glands in the area over their eyes to remove salt from the blood, thus permitting them to drink seawater and have a degree of freedom from land.

Marine birds, like their land counterparts, are light-weight and well insulated, but their most unique and remarkable evolutionary developments are in those areas relating directly to flight. Sustained flying over water permits few navigational or aerodynamic miscalculations. The wings of flying marine birds are long, pointed, and cupped underneath to support a generally large and muscular organism for long periods of time. Interestingly, they are used more for soaring and gliding than for flapping through a wide angle. The shape of the wings allows for high lift and low drag at high speeds.

Due to the rigorous nature of the marine environment, it is to a marine bird's advantage to be generally larger and stronger than land birds. The large size allows retention of heat, efficient aerodynamics, and the ability to fly strongly across great expanses of ocean. As they migrate or search for food, marine birds often fly astonishingly long distances in relatively short periods of time. The large albatrosses, for example, may remain aloft for many weeks or even months while moving over the Pacific, sometimes flying continuously eastward around the world in the West Wind Drift zone north of Antarctica. Arctic terns routinely complete an annual migration of over 40,225 km (25,000 mi)!

Not all marine birds fly. Penguins (order Sphenisciformes), for example, do not fly except through water. They are remarkably maneuverable in pursuit of their prey, fish. These familiar, short, heavy birds of the Southern Hemisphere can be as large as 1 m (33 in) tall and weigh as much as 7 kg (16 lb). There are 18 species of penguins, none native to the Northern Hemisphere.

The variety of oceanic birds is truly impressive. They range in shape and habit from the large pelican, which feeds by diving on its prey and entrapping it in huge gular pouches, to the frigate birds of the South Pacific, which feed by diving aggressively on other birds returning from a day's fishing. This dive, coupled with a loud hissing or screaming noise, causes the smaller birds to drop their catch which is then caught by the frigate bird. Gulls are plentiful on most temperate and tropical coasts and are often marvelously maneuverable flyers. All are interesting and attractive to the amateur bird watcher, and all share a distinct reptilian ancestry. It may seem odd, but the dinosaurs never really became extinct – they are alive today, and we call them birds.

Learning Objectives

After completing the reading assignment and viewing the program, the student will be able to:

☐ Recognize the common ancestry of reptiles and birds.

☐ List and describe the four main forms of living marine reptiles.

☐ Describe some of the adaptations of marine birds for long flights over water.

☐ Compare the behavior and habits of a number of oceanic birds.

Key Terms and Phrases

Mesozoic era The era of geologic time from 230 million years ago to 62 million years ago. The Jurassic period within this era, the middle of which was around 150 million years ago, saw the height of the domination of reptiles and the rise of birds.

Indo-Pacific Southwestern Pacific north of the equator.

Galápagos Islands off the coast of Ecuador, on the equator.

Sea turtles Oceanic members of the chordate order Chelonia. Some are very large with nonretractable heads and flippers.

Sea snakes Oceanic members of the chordate order Squamata. All sea snakes are poisonous.

Marine iguanas Oceanic members of the chordate order Squamata, suborder Lacertilia. They live only in the Galápagos Islands and are a remnant of ancient populations.

Oceanic birds Predominantly oceanic members of the chordate class Aves.

Before Viewing

☐ In Garrison read "Reptiles" beginning on page 400, and "Birds" beginning on page 401; in Ingmanson & Wallace read "Reptiles" beginning on page 316, and "Birds" pages 318–322.

☐ Watch Program 17: "Reptiles and Birds."

After Viewing

Reread the text section, being sure you understand the key terms and phrases, and complete the following exercises:

1. If marine reptiles were so advanced in the Jurassic period, why are they so much less successful today?

2. How can birds find their way? That is, how can they fly at night with no light, and how can they find their way from anywhere in the world to their own small nesting island?

3. What evidence can you suggest to connect dinosaurs to birds?

4. How do marine birds stay warm? Most productive marine environments are quite cold, and birds have the highest constant internal body temperatures of any animal.

Answer Key for Exercises

1. Difficult question! Many people have pondered for many years about this one, but here are some current speculations. (a) Climate changes may have made the environment colder, and the cold-blooded reptiles could not continue to be successful in the new climate; or (b) the emergence of new and successful mammals, especially the smaller ones that ate reptile eggs, posed a threat to the population of reptiles that eventually led to their demise; or (c) there was some sort of cosmic catastrophe that showered the earth with radiation, increased the mutation rate, and caused a simultaneous die-off of reptiles and increase in mammal populations; or (d) perhaps a combination of some or all of these factors. No one yet knows for sure, but it is interesting to note that after about 50 million years of dominance, the reign of the reptiles did come to a very abrupt end at about the same time as the rise of mammals.

2. Another difficult one. Some marine birds, notably albatrosses, can fly for extended periods without landing on water or shore. Not only that, but they fly in some of the most violent weather on earth near Antarctica. How they maintain altitude and fly

under these conditions is not understood. Long-range navigation has been studied, and it appears that birds use (among other things) celestial navigation to find their way to sea. The birds evidently have some sort of "star chart" designed into their brains, and along with a time sense (like a human chronometer) can find their location on the surface of the earth. Knowing their current location and knowing where they wish to go, they can complete the journey. Albatrosses have been transported over long distances by aircraft, released, and have turned up on their nests in surprisingly short times. Research at Cornell University has shown that navigating birds can use sun angles, time of day, magnetic forces, and even polarized light to assist in navigation. We have no full understanding of how bird navigation operates.

3. Recent discoveries suggest that some of the smaller, fast-running dinosaurs were not slow-witted, cold-blooded, cumbersome animals. Analysis of the system of bone canals within the fossils of these small dinosaurs suggests a surprisingly birdlike arrangement. Also, some of the bone structure and cranium design of these small and fleet dinosaurs is rather prebirdlike. The fossils of *Archaeopteryx* that have been discovered show a bony structure more similar to a delicate small dinosaur than to a modern bird, although the presence of feathers and other features on *Archaeopteryx* do place it clearly within the bird category. Much work remains to be done in this fascinating area of study. The book by Wilford listed in the Supplemental Reading for this chapter discusses the current state of research.

4. The flying birds use an extraordinary form of feather called "down," which may be familiar to you through its use in insulation for jackets and sleeping bags. Down has one of the highest "R" values (insulation value) known for any substance, synthetic or natural. As long as it is protected and kept dry by water-repellent outer feathers, the millions of tiny air spaces within down can insulate the bird from the bitter cold air through which it may be flying. One-half inch (about 1 cm) of down has the same insulating capability as 3.6 m (12 ft) of

construction brick! But down works only when it is dry, as any wet backpacker can attest. Diving birds and birds that swim for their food cannot make effective use of down for insulation, and they therefore substitute blubber. This fat is not nearly as efficient for insulation, so thicknesses must be greater. This, of course, means additional weight. For penguins the weight is not much of a problem, but for birds that dive *and* fly (puffins, for example), the penalty can be great. Puffins seem to struggle mightily to fly the few hundred feet from the water to their nesting cliffs. They have a thin layer of fat beneath a shield of closely packed water-resistant feathers. The feathers provide some insulation, though not as much as true down would give. So, warmth is a trade-off between mobility and where the food is.

Optional Activities

1. Bird watching is a great joy for many professional and amateur biologists. There are many excellent bird identification books, notably those by R. T. Peterson and by the Audubon Society. One of these books and a pair of binoculars can teach you a lot about the world of birds.

2. Many museums in the coastal areas of the United States have sea bird collections. If you have been lumping all gull-like birds together under the term *sea gull*, you have a surprise awaiting you.

3. The reptile house of most zoos has turtles and perhaps even sea snakes for you to observe.

Self-Test

(More than one answer may be correct.)

1. Reptiles were a dominant form of life on the earth during the _____ era, _____ years ago.
   a. Mesozoic; 2 million
   b. Jurassic; 2 million
   c. Mesozoic; 200 million
   d. Jurassic; 200 million
   e. Mesozoic; 150 million

2. All of these statements about marine turtles are true *except*:
   a. All are in danger of extinction.
   b. They have a strong homing instinct.
   c. Their front limbs are flattened.
   d. Their heads retract into their shells.
   e. Living species may exceed 10 ft in length.

3. Which of these groups does *not* have a strong homing instinct?
   a. marine birds
   b. marine turtles
   c. sea snakes
   d. marine iguanas
   e. marine crocodiles

4. The wings of the most "severely oceanic" birds (birds spending nearly all of their life span aloft, over the ocean) tend to be
   a. short and blunt
   b. short and pointed
   c. long, thick, and pointed
   d. long, thin, and pointed
   e. long, thin, and blunt

5. For navigation purposes, marine birds may use
   a. celestial charts built into the brain
   b. time-sensing abilities
   c. polarized light
   d. wind direction and sun direction information
   e. all of these, probably

6. Marine birds tend to be
   a. smaller than land birds
   b. larger than land birds
   c. smaller but stronger than land birds
   d. larger and stronger than land birds
   e. smaller and weaker than land birds

7. Marine birds eliminate salt by
   a. traveling to land to drink fresh water
   b. using salt glands to extract salt from the blood
   c. manufacturing fresh water from seawater
   d. extracting fresh water from their prey
   e. storing fresh water from rains within their feathers

8. Which of the following statements describes the internal temperature of oceanic birds?
   a. It is relatively low, nearly that of the surrounding air or water.
   b. It is high, but the internal metabolism of marine birds is so great that no special insulation need be provided.
   c. It is quite high, and the temperature is maintained by the use of down for insulation in all marine birds.
   d. It is quite high, with down and blubber being used for insulation as appropriate to the bird's lifestyle.
   e. It is quite high, with blubber being used in all birds, even those that fly for great distances in migrations.

Supplemental Reading

Bakker, R., "Dinosaur Renaissance." *Scientific American* (1975).

Carr, A., "The Navigation of the Green Turtle." *Scientific American* (1965).

Nelson, B., *Seabirds*. New York: Addison-Wesley, 1979.

Smith, N., "How Birds Fly." *National Wildlife*, vol. 13 (1976), 17.

Tickell, W. L. N., "The Great Albatrosses." *Scientific American*, vol. 223, no. 5 (1970), 84–95.

Wilford, J. N., *The Riddle of the Dinosaur*. New York: Knopf, 1986. (Part II covers the evolution of birds.)

Lesson Eighteen   Mammals: Seals and Otters

Overview

Marine mammals have captured public attention at oceanariums, on television, and in the popular press. The attractiveness and possible intelligence of small whales, the endearing performances of sea lions in circuses, the horror of the continuing slaughter of the world's largest animals – all have thrust this group of sea creatures into the contemporary consciousness.

However, despite the attention and fascination, marine mammals are in danger of extinction. Surely it seems ironic that a group that has come so far so quickly (in evolutionary terms) could simply cease to exist within a hundred years due to the shortsighted greed of the human species.

Mammals did not exist in the sea 50 or 60 million years ago. Then in a comparatively short period of time some terrestrial mammals left their four-footed land existence behind as environmental factors selected for characteristics favoring a marine existence. These environmental pressures resulted in the radical physiological and anatomical changes that have adapted them to the marine environment. The most important of these environmental factors was almost certainly increased opportunity for feeding. As a group, these marine mammals have given rise to the largest animals ever to live on this planet. In fact, the largest whales considerably exceed the mass and length of even the greatest of the dinosaurs.

There are three main groups of marine mammals, and each group has evolved independently from different terrestrial mammalian stock. The order Pinnipedia includes seals, sea lions, and walruses; the order Sirenia includes dugongs, manatees, and sea cows; and the order Cetacea contains whales. The order Cetacea will be covered in Lesson 19.

The pinnipeds are included within a larger group, often termed a "superorder," called Carnivora. This large group also includes otters and polar bears.

All marine mammals share a number of common features. Obviously, the shape of their limbs and body have been modified to assist them in moving efficiently through the water. In addition, good thermal insulation prevents excessive heat loss in the cold environment. Modifications to the respiratory system allow deep repetitive diving, and certain osmoregulatory adaptations free the animals from any dependence on fresh water.

The Pinnipedia (name means "feather-feet") is divided into three large families. The sea lions, members of the first pinniped family, the Otariidae, have small but obvious ears and long flexible necks, independent and mobile hind limbs that permit substantial movement on land, nostrils at the tip of the snout, and a dense and soft undercoat protected by long and coarse guard hairs. Northern fur seals, which are closely related to sea lions, may reach 2.2 m (7.5 ft) in length and weigh up to 318 kg (700 lb). Stellar sea lion males may reach over 4 m (13 ft) and weigh 1,000 kg (2,200 lb)! Females of the species are smaller.

Sea lions breed in rookeries on beaches where the dominant males will fight to protect their harem and territory. Females give birth on land to one pup, and mortality among young sea lions in the crowded rookeries is surprisingly high due to starvation, disease, or crushing. Estimates vary, but even after the young leave the rookery, some 50 percent are lost at sea through the first winter before the herd returns to the colony site.

Some Otariids in rookeries off the California coast may be suffering from the effects of contamination by chlorinated hydrocarbon pesticides used in agriculture. These chemicals, which wash off the land into the ocean, appear to be magnified through the food web, building up in the fat reserves of the animals. During those periods when females are ashore, living off stored fat and nursing their newly born young, the pesticides may cause behavior changes in both the adults and the young. Some of these changes seem to affect maternal behavior, resulting in abandonment or violence directed toward pups. Calcium deposition and metabolism in the growing pups can also be affected negatively. In addition, the proper action of the immune systems, essential for fighting off infection, is suppressed.

In the past few years the most offensive of these pesticides have been banned, and for awhile the situation in the rookeries off California began returning to normal. Unfortunately, there are recent indications that "bootleg" DDT has been illegally reintroduced into California agriculture. If this is true, the sea lions will probably undergo another anxious and dangerous period of human interference.

The second pinniped family, Phocidae, is comprised of seals. Seals differ from sea lions in that they have no external ears, possess a smooth coarse hair with no underfur, and have a hind-limb geometry that restricts their ability to move freely on land. Their foreflippers are smaller than those of the eared sea lions. The size attained by seals is quite impressive, and the northern elephant seal reaches 4.8 m (16 ft) long and 2,270 kg (5,000 lb).

Breeding behavior in seals is variable. The elephant seal, for example, breeds in large colonies to which they return after many months of hunting. Other varieties, such as the frequently encountered harbor seal, do not congregate in large numbers, breed in colonies, or maintain harems.

The Harp seal, a species found in relatively great abundance on ice floes off the northeast coast of Canada in the Labrador Sea and in the Barents Sea, has been in the news continuously during the last decade. The controversy concerns the "harvest" of Harp seal pelts. Newborn pups of dense and pure white fur have been systematically slaughtered each spring for their pelts to clothe the fashion-conscious of Europe and North America. Because these seals are so attractive and helpless, much effort has been made to stop their slaughter. In fact, they are not an endangered species, and the Canadian government carefully monitors the numbers of

animals taken. Still the basic question remains regarding the relative value of a seal pup's pelt made into a coat.

The third pinniped family, Odobenidae, includes only walruses. These very large animals are limited to arctic waters in the Atlantic and Pacific Oceans. They have hind limbs similar to those of sea lions, but are earless. Their most identifying feature is the set of canine teeth that has been modified through the years into large tusks for digging the shellfish they eat. Their other teeth are peg-shaped and are used for crushing. These animals are becoming very rare, and population estimates place the world number of walruses at between 170,000 and 190,000 animals.

Perhaps the least-known and most unexpected marine mammals are the members of the order Sirenia. These improbable and gentle vegetarians were named for their fanciful resemblance to the mermaids encountered by Ulysses on his return from the Trojan Wars. They live in bays and estuaries in the southeastern United States, on both sides of Africa, and in the Indo-Pacific, but not in open sea. They exhibit little fear of man, form family groups, and show considerable affection and concern for their offspring. They feed on water plants. Sirenians were introduced into this country in an attempt to control water hyacinths in canals and waterways in the Deep South.

All species of sirenia are in grave danger of extinction; only 10,000 sirenians may be left in the world. The West Coast of the United States was once the range of very large sirenians called Stellar's sea cows. By 1741, thirty years after the time of their discovery, hunters and whalers had exterminated the species. A similar fate almost befell the West Coast's sea otter.

Sea otters have the unfortunate distinction of possessing the most valuable fur of any animal. The dark brown underfur is extraordinarily rich and dense. Although trade in sea otter fur dates back to at least 1786 under the Spanish, it includes traders of many nationalities. In fact the otter fur trade became the most important industry of the California coast and was a factor in the settlement of the Pacific coast of North America. By 1900 otters were considered extinct along the California coast because of the fur slaughter. In the 1930s, however, a small population was discovered near the mouth of Bixby Creek, south of Carmel on the central California coast, and today the otter's range has expanded along much of the Pacific coast of the United States.

Sea otters are weasel-like mammals closely related to the river otter. They grow to nearly 1.5 m (5 ft), may weigh 36 kg (80 lb), and can swim at 5–10 mph. They live for up to 20 years and seem to derive great pleasure from play activity throughout their lives. They forage for urchins, abalone, mussels, crabs, and other tidbits in relatively shallow water, rarely diving to depths of more than 18 m (60 ft). Their food is frequently consumed leisurely as they float gently on their backs. They love surfing and probably consider water at rest to be pretty dull. Otters are *very* entertaining to watch!

Though legally protected, otters are still under great pressure. They compete with abalone fishermen, rob lobster traps, and get into so much mischief they are not equally welcomed by all people. Many otters are consumed by killer whales, and some are consumed by sharks.

The polar bear is found throughout the high arctic. Camouflaged by its white fur, this mammal is a swift and wide-ranging traveler. It captures and consumes primarily pinnipeds, but will supplement its diet with other marine food (fish and seaweeds) and even land-based food (caribou and grasses) when they are available. Polar bears are very large and dangerous animals, weighing up to 720 kg (1,600 lb). Their height can range up to 1.6 m (5.3 ft) at the shoulder! Migrations across pack ice by walking and swimming to follow seal populations are facilitated by hairy soles on their feet, which insulate from the cold.

The existence of marine mammals is in itself astonishing. Although they definitely show evidence of their land-mammal heritage, it is hard to imagine how so diverse a set of organisms evolved in so compact a time span. However, they did, and humanity is the better for it.

Learning Objectives

After completing the reading assignment and viewing the program, the student will be able to:

☐ Recognize the ancestry of marine mammals and discuss their derivation from land mammals through a span of 60 million years.

☐ List the three main groups of marine mammals and compare their basic characteristics.

☐ Differentiate between seals and sea lions.

☐ Understand that all marine mammals are in danger of extinction.

Key Terms and Phrases

Mammals Warm-blooded organisms of the phylum Chordata. These animals give birth to living young, suckle the young with milk, have hair, and possess the most advanced brains in the animal kingdom. They are currently the dominant large organisms on the earth.

Marine mammals Specifically adapted mammals designed for a marine existence. They have a number of unique features and have evolved in the last 60 million years from land forms.

Pinnipedia Marine mammals; specifically seals, sea lions, and walruses.

Sirenia Marine mammals; specifically dugongs, manatees, and sea cows.

Cetacea Marine mammals; specifically whales, including dolphins, porpoises, and larger forms.

Fissipedia Mammals in the superorder Carnivora living in the oceanic environment; specifically sea otters and polar bears. These are not true marine mammals in the same sense as cetaceans, sirenians, and pinnipeds.

Before Viewing

☐ Read "Marine Mammals" on pages 403–410 of Garrison, or "Mammals," pages 332–335, 429–431 of Ingmanson & Wallace.

☐ Watch Program 18: "Mammals: Seals and Otters."

After Viewing

Reread the text sections, being sure you understand the key terms and phrases, and complete the following exercises:

1. Why did the evolution of marine mammals proceed at such an apparently fast pace? Sixty million years does not seem like a very long time, relatively speaking, to derive such elaborately adapted creatures.

2. Large numbers of seals, sea lions, and even otters continue to be killed each year. Can't anything be done to stop this? Should anything be done?

Answer Key for Exercises

1. The evolution of organisms on earth occurs at a deliberate pace and can never be hurried. It is directed by environmental conditions and is simply the response of living things to changes. Marine mammals arose from different terrestrial stocks, and the first forms spent only a little time in the sea, perhaps in enclosed bays and estuaries. The feeding situation in these locations was probably very good, and the generations of organisms to follow spent progressively more time in the water as aquatic feeding structures and behavior were "selected for" by the environment. Eventually the organisms changed sufficiently to abandon their land existence altogether. The fossil record suggests that this occurred through about 60 million years and resulted in the organisms we see today. The rate of evolution was not fast or slow, but rather was appropriate to the conditions as they changed and the organisms as they developed.

2. Many groups, including large numbers of organizations in the United States, Canada, and Great Britain, have tried for years to have the killing of all marine mammals stopped. The sight in the spring of the slaughter of the small, cuddly, white, newborn Canadian Harp seals is an anguishing vision for most marine-oriented people and friends of animals in general. But some populations of seals, sea lions, and other marine mammals may be capable of withstanding considerable thinning without population stress. Other populations, notably the great whales and otters, may *never* recover from what was thought at one time to be adequate kills. Who should make these decisions, and upon what data these decisions should be based, are questions that are currently entangled in controversy. As in so many other cases, no simple answers are forthcoming. We must balance our needs against the capability of the populations to reproduce themselves.

Optional Activities

1. Oceanariums have all sorts of opportunities for the public to see marine mammals in action. It would be useful to read something about the animals you expect to encounter before seeing them.

2. One of the most effective organizations working for the preservation of marine mammals is the American Cetacean Society, a nonprofit organization. You may wish to contact them for information. Their address is P. O. Box 1391, San Pedro, California 90733.

Self-Test

(More than one answer may be correct.)

1. Marine mammals began to evolve distinctly from land mammals about _____ million years ago.
   a. 2
   b. 20
   c. 60
   d. 250
   e. 500

2. Which of these organism groups has *not* been traditionally classified as a typical marine mammal?
   a. Fissipedia
   b. Sirenia
   c. Cetacea
   d. Pinnipedia
   e. none of the above

3. Which of these is *not* a pinniped?
   a. seal
   b. walrus
   c. dolphin
   d. sea lion
   e. sea otter

4. Which of the following is (are) characteristic of a sea lion?
   a. It is a pinniped.
   b. It has visible external ears.
   c. It has independent, motile hind limbs.
   d. It forms harems and territories.
   e. all of the above

5. The most valuable fur in the world belongs to members of what group?
   a. Pinnipedia
   b. Cetacea
   c. Fissipedia
   d. Sirenia
   e. Phocidae

6. The largest pinnipeds are the
   a. walruses
   b. sea otters
   c. whales
   d. sea lions
   e. seals

7. Which of these statements is *not* true regarding sea otters?
   a. They were nearly extinct along the West Coast of North America by the 1930s.
   b. They produce a dense, rich fur.
   c. They are now strictly protected by law.
   d. They are no longer killed.
   e. They are easily found and fun to watch along the central California coast.

8. Which of these marine mammals does *not* engage in migratory and/or homing behavior?
   a. otters
   b. polar bears
   c. seals
   d. sea lions
   e. All of the above mammals migrate and return to a home base.

Supplemental Reading

Hedley, D. (ed.), *Marine Mammals,* Second Edition, revised. Seattle: Pacific Search Press, 1986. (A collection of papers and articles of general interest.)

Jensen, A. C., *Wildlife of the Oceans.* New York: Abrams, 1978.

Oceanus, vol. 21, no. 2 (1978). (The whole issue of this *Oceanus* is devoted to marine mammals.)

Orr, R., *Marine Mammals of California.* Berkeley: University of California Press, 1972.

Lesson Nineteen   Mammals: Whales

Overview

No person can fail to be impressed by underwater photographs of great whales swimming, or by the sight of a whale leaping from the water and returning with a spectacular splash, or by the number of whales that have fallen victim to exploitation by humans. Some whales have been found to sing intricate and beautiful songs; some can move through the water at seemingly impossible speeds. Whales are certainly the most dramatic creatures of the ocean, deserving of our study and protection.

Whales and porpoises belong to the mammalian order Cetacea. As a group, cetaceans exhibit nearly complete hair loss, possess a large brain with deep convolutions, are covered by a special skin that selectively "gives" to damp out laminar water flow irregularities, and are uniquely powerful among animals.

Being mammals, whales, of course, have a constant internal body temperature much higher than that of the surrounding water. To insulate the body, whales have a thick layer of greasy, buoyant fat called "blubber" beneath their skin. This layer of fat, combined with a marked reduction of surface capillaries, allows these animals to maintain a high temperature in a cold, productive sea. It is ironic that many of the largest species of the large, warm whales require so much food that they must live in the productive polar seas near permanent pack ice. Blubber helps make this possible.

The cetaceans are further divided into two groups, the suborder Odontoceti (toothed whales) and the suborder Mysticeti (baleen or great whales). Toothed whales may be the more familiar, particularly to those who visit oceanariums. In general, toothed whales with their powerful jaws are smaller than baleen whales.

Small toothed whales of the Odontoceti subclass function well within their environment, although estimates on precisely where these animals should be placed on the mammalian intelligence scale are still controversial. Small whales have the largest brain-to-body weight ratio of all animals, and the brain's surface area is proportionally large due to an extensive system of folds. Marine scientists who have worked with dolphins and porpoises for many years often develop the feeling that the animals manipulate their trainers and are capable of deep understanding. However, attempts to quantify the level of intelligence have not withstood repeated testing. Perhaps if current attempts to bridge the language gap between humans and dolphins are successful, we will learn more about their relative intelligence.

Toothed whales tend to be quite vocal, and many species use sonar to locate prey. In fact, the sonar system of small whales deserves special attention. After World War II and the invention of naval sonar (an acronym for *so*und *na*vigation and *r*anging), it became apparent that much of the ocean's background noise was coming from small whales. Also, most of the small whales seemed able to travel confidently through the water at very high speeds even when they could not see through the water for even a fraction of their safe stopping distance. Researchers put these facts together and hypothesized the existence of some analytical mechanism within the animals that allows them to "see" with reflected sound. Further evidence was supplied when zoologists found that small whales have a very large acoustic cranial nerve, a brain with extraordinarily large acoustical receptor areas, and inner ears especially adapted to the reception of very high frequencies.

In addition, whales have a membranous container of oil located in front of the sound-producing organ and beneath the skin at the front of the animals' heads. This reservoir, referred to as a "melon" in smaller odontocetes, is evidently used to focus sound emitted from the whale. It may also play a role in sound reception. The muscular control a whale can exert over the "melon" allows it to vary the sound patterns. A sound pattern can be broad and fan-shaped for general ranging, or very thin and narrow (a "pencil beam") for use in high-speed sonar scanning. Other whales may use the pencil-beam mode for even more spectacular purposes, as we shall soon see.

Tests with blindfolded dolphins in pool obstacle courses have also verified the sonar hypothesis. In fact, since the 1940s, much has been learned about the physics of underwater sound from work with small whales.

The largest toothed whale and the possessor of the largest brain ever to evolve on the earth is the sperm whale. These whales are 18 m (60 ft) long and can dive to greater depths than any other air-breathing animal. A sperm whale was once found entangled in an oceanic telephone cable at a depth of 1,134 m (3,720 ft). In the sperm whale the reservoir of oil in its head is called a "case" and may weigh up to 909 kg (2,000 lb, or 1 ton). It is believed that sperm whales use this sonar scanning system to hunt prey.

Marine biologists have wondered for years how a sperm whale can catch the deep-water squid that make up most of its diet. The squid would seem to have a real advantage. It can swim faster than sperm whales, and has a biting beak and ten tentacles. (Sperm whales don't even have teeth until they reach sexual maturity at about 20 years of age.) Recent research by Dr. Kenneth Norris and others indicates that the sperm whale can confuse, stun, or even kill their adversary by blasts of intense sound from their sound-making machinery! The action begins at the surface with the sperm whale scanning the depths with sonar. If a squid is detected, the sperm whale changes its sound-making mode from "scan" to "stun" and begins its descent. As the sperm whale nears the squid, a "stun" sound is generated within the head of the sperm whale, focused by the oil sac in the head, and aimed at the squid while the whale is still descending. The sound that is generated is unbelievably intense, up to a calculated maximum of 260 dB at its center of focus. To give you some basis for comparison, a jet airliner at the end of a runway will generate about 180 dB.

Obviously more research will have to be done in this area, but much evidence suggests that this strategy is indeed being employed by these animals. Some of the smaller odontocetes (dolphins) may also use this sort of aggressive sound. Bursts of sound have been measured at 228.6 dB from bottlenose dolphins. Some cetacean biologists are not convinced that whales actually use sound in this way, but if they do, the term "mindblowing" takes on new meaning in an oceanic sense.

The second cetacean group, suborder Mysticeti, is more highly evolved and contains generally larger species than the odontocetes. This group contains the blue whale, the largest animal ever to live on the earth. Blues are animals of superlatives. They reach over 30 m (100 ft) in length, and can weigh 82,000 kg (90 tons). The power they generate as they move through the water has been measured in excess of 373 kilowatts (500 horsepower). Blue whales perhaps communicate with each other across entire oceans at frequencies far below the limit of human hearing (10–20 hertz). They eat small planktonic arthropods called krill (genus *Euphausia*) and are in grave danger of extinction.

Blue whales, like other members of suborder Mysticeti, feed by filtering plankton from the water with plates of baleen. This bristly, bushy, and fibrous substance is set in the jaws in overlapping plates. The animal takes a mouthful of water, closes the jaws, raises its tongue to expel the water, and swallows the plankton trapped on the inner surfaces of the baleen. A large blue whale swims at a nearly continuous speed of 10–12 nautical mph, and requires an estimated 780,000 calories per day for propulsion. Just to maintain body metabolism this animal requires another 230,000 calories per day for a total estimated daily need of 1,010,000 calories. Krill contains approximately 460 calories per pound, so blue whales must eat about 3 tons of krill per day to survive.

Prior to whaling there were around one-half million blue whales in the Southern Hemisphere. Estimates of today's population place the number of blue whales at one thousand. Imagine the oceanic imbalances this huge reduction in the whale population has caused.

Perhaps the whale that has received the most attention recently is the celebrated and melodious humpback. These whales sing a loud, precise, and repetitive melody during the half-year they are involved in mating and calving in tropical waters. They do not sing during the other half-year while they are feeding in the food-rich polar ocean. All members of the same population group sing the same song which subtly evolves over the six months it is in use. Then, after six months of silence, the song is considerably changed and all members of the population adopt the new song. The song is always rigidly structured with each whale in the large group singing an almost identical version. The song evidently serves as an information-carrying and identification tool. Though researchers are divided on its significance, marine biologists continue to be enthralled with the beauty and complexity of these noises. As whale songs are studied in more detail, they are found to be even more intricate than was first thought.

Whale songs are probably the first oceanic noises to capture the imagination of the general public. A brief recording of humpback whale songs was included by Carl Sagan in the golden phonograph record on board *Voyager 10* in 1977 (*Voyager* is now beyond the solar system), and some humpback whale songs have been recorded for commercial sale. A list of recommended recordings is included at the end of this lesson.

As noted on page 453 of Garrison, or on page 406 of Ingmanson & Wallace, except for suspiciously large numbers of whales taken for "scientific" purposes, the capture of large whales has ended. Unfortunately the slaughter of small whales (dolphins) continues in association with tuna fishing. Spinner dolphins and spotted dolphins often accompany yellowfin tuna, the tuna preferred by most Americans. The tuna/dolphin problem is doubly troublesome in that no use whatsoever is made of the small whales ensnared and drowned in the purse seines. Modern purse seine fishing methods, in fact, depend in part on these mammals as markers for the fish. In 1987, 140,000 dolphins were killed this way worldwide, some 26,000 by U.S. fishermen. These numbers will perhaps decline further in the near future – the U.S. Congress voted in October, 1988, to ban importation of tuna not caught in accordance with new U.S. methods designed to reduce the kill. Consumers can purchase albacore tuna, a species not associated with dolphins.

Learning Objectives

After completing the reading assignment and viewing the program, the student will be able to:

☐ Distinguish between cetaceans of suborder Odontoceti and suborder Mysticeti.

☐ Explain, in general terms, the functions of odontocete sonar.

☐ List a number of important cetaceans and give distinguishing characteristics for each.

☐ Know some uses for whale sounds other than sonar.

Key Terms and Phrases

Cetacea Marine members of the chordate class Mammalia. Order Cetacea is divided into two suborders – Odontoceti and Mysticeti.

Odontoceti The cetacean suborder of toothed whales. Example: killer whale, sperm whale.

Mysticeti Cetacean suborder of baleen whales. Examples: gray whale, blue whale (synonyms: great whales, baleen whales).

Baleen Fibrous plates of filtering material found in the mouths of whales of the suborder Mysticeti.

Sonar Acronym for *so*und *na*vigation and *r*anging. Sonar is the method by which underwater sound is used to discover position and targets.

Song Not a set of random noises, but an organized and repeated system of sounds made for a specific purpose.

Before Viewing

☐ Review "Marine Mammals" on pages 403–410 of Garrison, and read pages 452–454. If you are using Ingmanson & Wallace, read "Cetaceans" on pages 325–329, and reread "Mammals" on pages 411–413.

☐ Watch Program 19: "Mammals: Whales."

After Viewing

Reread the text sections, being sure you understand the key terms and phrases, and complete the following exercises:

1. Baleen whales are the largest animals on the earth, yet they eat small plankton animals. Why did such big animals evolve to eat such tiny food?

2. Blue whales can evidently communicate over entire oceans. How is this possible?

3. What is the difference between a dolphin and a porpoise?

4. Much of the mortality of small whales in tuna fishing is due to drowning. How do these maneuverable animals drown?

5. Would it be possible to farm great whales? Is this a possible way of saving them from extinction?

6. It would appear that a small, fish-sized marine mammal would be a reasonable evolutionary outcome. The combination of muscle physiology, intelligence, anatomy, and jaw structure makes the logic of compact marine mammals very appealing. But there is none smaller than about 2 m (6.5 ft) long. Why?

7. If sperm whales can make such intensely loud sounds, perhaps capable of stunning or killing large and powerful animals like squid, how does the sperm whale itself survive the production of the intense burst of sound? What makes the sound in the first place?

Answer Key for Exercises

1. The answer involves feeding efficiency and food chains. Much energy is lost as flesh passes up toward a top consumer in a food chain. As little fish eat plankton, larger fish eat the little fish, and humans eat the larger fish, efficiency goes down. In eating tuna, we are far removed from the original phytoplankton that captured solar energy in the first place. Solar energy powers all organisms.

   Great whales eat from near the base of the food chain. In this way they can take in huge quantities of energy without requiring the support of middle links in the food chain. The situation is analogous to a buyer's desire to purchase an item directly from the producer rather than having to go through two or three middlemen, each of whom adds his own profit to the ultimate price.

2. The trick here is in the very low frequency of the sound used by the blue whales. The sound is so low in frequency that it was not discovered until recently. Sound at these frequencies travels easily in the ocean and is diminished only slightly after hundreds of miles of passage. Sounds often travel for vast distances in thermocline-formed "sound pipe" channels near the surface of the sea. Blues may take advantage of these channels for communication over planetary distances. Submariners have known about such channels for years but the whales have evidently been using them for eons. No one yet knows how blue whales generate this powerful and awesome sound or what use they make of it.

3. There is much confusion over these terms. There is a dolphin fish, which is also called mahi-mahi, prized for its fighting ability. Although this fish has the same common name as the mammalian dolphin, it is not a mammal. The small dolphin mammal can be distinguished from the porpoise by its long, bottle-shaped snout. The head of the porpoise tapers abruptly to the jaw line. Killer whales are porpoises.

4. Purse seine nets, when nearly closed, are shaped like an inverted light bulb – a large round shape below a thinner neck area. Dolphins and porpoises, like all cetacea, are air-breathing mammals. When they cannot reach the top of the net for air, they drown.

   When these animals are badly frightened, there is some evidence that they will not jump over the lip of the net. In a life-threatening situation they may choose, as humans might, to stay in an environment they fully understand, rather than to abandon it for one in which they spend only short amounts of time. Developments in the tuna industry that provide back-down directions to boat captains and special "Medina strip" nets have alleviated much of the mammal kill associated with the fishery.

5. The space requirements alone are boggling. The volume of planktonic food required would be very, very large. The management of such resources and the successful breeding of these animals in captivity are unknown quantities. It might conceivably be possible, but it would be prohibitively expensive. Maybe if we just leave them alone they will rebound. California's gray whales did just that.

6. It is an interesting question! The logic of the answer revolves around the matter of heat loss to the water. The smaller the organism, the less favorable its surface-to-volume ratio for the retention of heat. Mammals, by reason of their physiology, require not only a steady internal body temperature but also a rather high internal body temperature. The only organisms with higher internal temperatures are the birds. As a marine mammal is "scaled down," it tends to lose an even greater percentage of its internal heat through the skin. Small marine mammals could not, of course, retain the thick layers of blubber found in their larger cousins. So, there are no small cat-sized marine mammals because they would lose heat to their environment too rapidly. The fish with an occasional higher-than-ambient internal temperature (such as tuna) do not maintain as great a relative temperature difference between their muscles and the surrounding water as do the marine mammals. Their heat loss is consequently considerably less, so they can operate successfully.

7. The sound is generated within a part of the breathing apparatus at the rear of the "case" near the cranium of the whale. It is believed that sound bounces off the cranium, moves forward, enters the case, and is focused there toward the necessary narrow-beam shape. The beam leaving the head is still relatively broad and comes to a point at some distance ahead of the whale, where the sound energy is concentrated.

To the old-time whalers, the oil in the case was the most prized part of any whale. Even today the oil is sought (usually on the black market) as a marine lubricant of great value. The whalers gave the sperm whale its name because they mistook this oil for seminal fluid, apparently overlooking the fact that the oil is present in both sexes of sperm whales!

Optional Activities

1. The California gray whale comes close enough to the shore to be seen during its migrations toward the Baja California calving grounds each winter. Thousands of people go on whale-watching trips to see these animals. If possible, you might take advantage of such trips, which leave regularly from San Pedro, Newport, Dana Point, and San Diego harbors. The American Cetacean Society can provide additional information. Their address is P.O. Box 1391, San Pedro, California 90733; telephone number (310) 548–6279.

2. Many whales frequent the coastal areas. When you are near the beach, keep your eyes open for cetaceans.

3. Compact discs and audiocassettes of whale sounds and songs are available. One of the best, "Deep Voices," contains the newly recorded low-frequency noises of the blue whale. If you have exceptionally good audio equipment, you will especially enjoy these low frequency sounds. The neighbors will think it is an earthquake. (See the end of this lesson for other listings.)

4. Again, oceanariums exhibit a variety of small whales. Sea World in San Diego even had a mysticete whale for a time (Gigi, the gray whale), but she proved too difficult to keep in captivity. You may wish to visit one of these facilities to see these animals.

Self-Test

1. Which of these characteristics is *not* applicable to all Cetacea?
   a. almost complete hair loss
   b. large, deeply convoluted brain
   c. air-breathing
   d. teeth in powerful jaws
   e. skin that decreases the friction of water flow at middle and high speeds

2. These whales are also known as whalebone or baleen whales:
   a. whales of suborder Odontoceti
   b. whales of suborder Mysticeti
   c. all whales
   d. all whales and whale sharks combined
   e. none of these

3. Sonar is used by
   a. all whales
   b. toothed whales, primarily
   c. baleen whales, primarily
   d. the navy and bats, but not whales
   e. none of these

4. Which of the following statements applies (apply) to the largest whales?
   a. They are members of the mysticete group.
   b. They are larger than any dinosaurs.
   c. They are capable of eating only small plankton as food.
   d. They are capable of communicating over large distances.
   e. all of the above

5. The largest whale is the
   a. sei whale
   b. gray whale
   c. blue whale
   d. killer whale
   e. minke whale

6. Which of the following statements describe(s) the smallest whales?
   a. They are members of the odontocete group.
   b. They catch their prey with teeth.
   c. They use sonar.
   d. They appear to be quite intelligent.
   e. all of the above

7. Like other marine mammals, whales have evolved from land ancestors.
   a. True. It took them about 60 million years.
   b. True. It took them over 200 million years.
   c. True. It took them about 2 million years.
   d. False. They evolved in the ocean and never had land ancestors.
   e. False. They evolved from fish.

8. The U.S. continues to kill what marine mammals for commercial purposes?
   a. great whales
   b. seals
   c. porpoises
   d. sea otters
   e. none of these

9. Blubber can be described by which of the following?
   a. It is a thick fatty material found in marine mammals.
   b. It is quite oily in appearance.
   c. It insulates the whale's body heat.
   d. It makes the whale buoyant since it has no swim bladder.
   e. all of these

10. Which whales have been found to "sing"?
    a. grays
    b. sperms
    c. seis
    d. dolphins
    e. humpbacks

11. The largest toothed whale is the
    a. blue
    b. sperm
    c. gray
    d. killer
    e. finback

12. The body temperature of order Cetacea
    a. is identical in all species of the order
    b. never varies throughout the day in response to environmental constraints
    c. is much higher than that of their surroundings
    d. is about 100°C
    e. none of these

13. Which whale is thought to stun or kill its prey using intense bursts of sound?
    a. narwhal
    b. sperm whale
    c. blue whale
    d. humpback whale
    e. beluga whale

14. The purpose of the "melon" or "case" in odontocetes appears to be
    a. to filter nutrient material
    b. reproduction
    c. to provide necessary weight for diving
    d. to prevent "the bends"
    e. to focus and manipulate sound

Supplemental Reading

Clarke, M. R., "Head of the Sperm Whale." *Scientific American* (1979), 128.

MacIntyre, J., *Mind in the Waters*. New York: Scribner's, 1974.

Minasian, S., *The World's Whales – The Complete Illustrated Guide*. Washington, D.C.: Smithsonian Books, 1984. (Extraordinary photographs of and concise information on nearly all of the world's whale species arranged taxonomically. A beautiful book.)

Oceanus, vol. 21, no. 2 (1978). (This whole issue is devoted to marine mammals.)

Roper, C. E. E., and K. J. Boss, "The Giant Squid." *Scientific American*, vol. 246, no. 4 (1982), 96.

Rudd, J. T., "The Blue Whale." *Scientific American*, vol. 195, no. 6, 46–65.

Sanderson, S. L., and R. Wassersug, "Suspension-Feeding Vertebrates." *Scientific American*, March 1990, 96–101. (Animals that can eat plankton grow in huge numbers or to enormous size.)

Tangley, L., "A Whale of a Bang!" *Science 84* (1984), 73–74. (An astonishing short article on the possible use of sonic bursts for hunting by odontocetes.)

Audio Recordings of Whale Sounds

"Deep Voices." Capitol Records ST-11598 (1977).

"Songs of the Humpback Whale." CRM recordings (1970). (Available from the American Cetacean Society, P.O. Box 1391, San Pedro, California 90733, (310) 548–6279.)

"And God Created Great Whales." A. Hovhannes, Andre Kostelanetz, and Columbia Symphony Orchestra. Columbia Records #M-30390 (1975).

"Sounds of Sea Animals." Folkways Science Series FX-6125 (1959). (This recording is particularly interesting because it contains slowed-down sounds of small whales enabling the listener to detect the reflected echo. Try to imagine, as you listen to this disk, how difficult it must be for a brain to construct a physical view of the world using just reflected sound!)

Lesson Twenty   Living Together

Overview

Newcomers to biology are often surprised to learn that more than half of the animal species known to science are not "free living." Most are actively involved in some sort of direct symbiotic relationship with at least one other life form. These relationships are often intricate and sometimes quite strange.

Symbiosis is the biologist's term for a close association between organisms. The word itself is derived from Greek meaning "with-life." The strength of the symbiotic bond is often so strong that one organism is totally dependent on the other. Symbiosis may be categorized as either direct or indirect.

There are three types of direct symbioses. In *mutualism,* as the name implies, both the symbiont and the host benefit from the relationship. True mutualism is rare in biology. For example, it was once assumed that the relationship between the pilot fish (remora) and the shark near which it swims was mutualistic. The pilot fish would guide the shark to a meal and in turn was permitted to dine on whatever scraps remained. We now know that the remora plays no role in the acquisition of prey by the shark, but merely helps itself to scraps after the shark has eaten its fill.

Many good examples of mutualism are observed in the ocean, however. One is the clownfish/anemone association. In this symbiosis the small, brightly colored clownfish (genus *Amphiprion*) nestles within the toxic tentacles of a sea anemone. The mechanism that permits the fish to do this without being stung is not fully understood, but marine biologists believe the fish makes use of an acclimatization system between the mucous secretions of its own skin and the stinging cell triggers of the anemone. In return for this protection, the fish feeds the anemone scraps of food and may even attract prey toward the anemone's reach.

Another example is the green alga that lives within the tissues of the common West Coast anemone *Anthopleura*. The anemone provides a secure place for the alga to live and a dependable source of carbon dioxide in return for the oxygen and carbohydrates produced by the alga. A similar relationship exists between the very large tropical clam *Tridacna* and its "live-in" algae.

Another "live-in" alga allows coral organisms to build gigantic tropical reefs. Without the interior resident alga known collectively as *zooxanthellae*, reef corals would be unable to deposit calcium and the marvelous tropical reefs we know today would not exist.

Perhaps the most striking examples of mutualism involve *cleaning symbioses*. In these relationships small cleaner organisms (usually fish or shrimp) establish a "cleaning station" to remove troublesome surface parasites from the skin, mouth, and gill coverings of larger animals (usually fish) visiting the site. The cleaner eats the parasites, so both animals benefit. The cleaned animals frequently defend the station and the cleaners from attack by other, less informed predators.

This idyllic situation seems almost too good to be true, and sometimes it is. One such flawed cleaning association involves the cleaner wrasse, genus *Labroides*, and its mimic, a sabre-toothed blenny, genus *Aspidontus*. When a larger fish (a red snapper or grouper, for example) approaches the cleaning station, it sees the familiar cleaner fish and relaxes, entering the "trancelike" state so common in cleaning customers. The big fish's mouth gapes open and its gill covers fan out. The cleaner wrasse then swims into the mouth, removing parasites and lodged particles of food, and even cleans the gills. Extremely sensitive eyes are groomed as well.

However, much to the larger animal's surprise, the sabre-toothed blenny, a fish almost indistinguishable from the cleaner wrasse, may also inhabit this cleaning station. When this impostor takes a big bite from the edge of a fin (or gill), the startled and pained large fish wheels quickly around only to find the offending mimic calmly relaxing and seemingly unaware of the source of the commotion. In order for this guise to succeed, the mimic must look and act like a legitimate cleaner wrasse. The impostor blenny, in fact, is so versatile a mimic that it can change color, as well as behavior, to match the situation. It looks and acts like a cleaner wrasse only when a likely victim is nearby!

In *commensalism*, another type of direct symbiosis, the symbiont benefits from the association while the host partner neither benefits nor is harmed. The pilot fish/shark relationship, mentioned above, is an example of commensalism. Before the clownfish was observed feeding the anemone, the clownfish/anemone association was thought to be commensalistic. Some of the most curious commensals are those that enter their partner's habitat, and after a period of growth, cannot escape. Small pea crabs (genus *Fabia*), for example, live in mussel shells, eating particles of food that are brought inside by the normal feeding and respiratory actions of the mussel. After a time, they grow so large that they cannot leave the mussel because they are larger than the gap between the shells.

The last direct symbiotic relationship is the most highly evolved and by far the most prevalent: *parasitism*. The parasite lives on or in the host and obtains food at the expense of the host. For obvious reasons, parasites do not usually kill their hosts, but they can seriously affect the host organism by reducing its feeding efficiency, depleting its food reserves, and lowering its resistance to disease. The host/parasite relationship is extraordinarily delicate, and the parasite must in some way be aware of the physical condition of the host in order to avoid taking so much energy from the host that the host is terminally weakened.

The most widely distributed and successful parasites are the round worms of the nematode group. Like most parasites, nematodes are quite species-specific in their selection of a host, The reason for this is the delicacy of the host/parasite relationship mentioned earlier. The complex biochemical feedback mechanisms used to inform the parasite that it may be overstressing the host are, by nature, limited to certain parasite/host pairs. If a parasite invades a host for which it is not specifically "programmed," either it will not survive or it will kill the host through overvigorous feeding.

More than one species of parasite, however, can infect a single host. In fact a significant percent of the weight of many fishes may be in nematode worms and other parasites. The parasitic burden of a normal, apparently healthy sea lion may exceed 5 lb and 20 species!

These three categories (mutualism, commensalism, and parasitism) form a sort of continuum in nature. There are few clearly mutualistic relationships that do not suggest at least a touch of commensalism, and few commensalistic relationships that do not border on parasitism.

Some marine biologists consider normal food webs to be a variety of symbiosis. These are called indirect symbioses or dependencies and are more general in nature than the direct symbioses described above. For example, many marine animals can capture and eat only very specific kinds of prey species. One particular species of copepod (a planktonic arthropod) has been discovered that can eat only one particular size phase of one species of diatom because of the configuration of the copepod's mouth parts. So specific a dependency is unusual. Most organisms are designed to eat a broader range of food species. These organisms exist within complex interrelationships known as food webs. Within food webs only about 10 percent of the mass of the food consumed is converted into mass within the consumer. Therefore, 10 pounds of small fish would be required to support one pound of tuna per year.

Although it is not formally included within the definition of the term, in an even broader sense all life on earth is functioning in symbiosis. Except for energy input from the sun and the radiation into space of waste heat, the planet is a closed system.

Learning Objectives

After completing the reading assignment and viewing the program, the student will be able to:

☐ Describe the relative abundance of free-living versus symbiotic species.

☐ Understand the origin of the word *symbiosis*.

☐ Contrast direct symbiosis with indirect symbiosis.

☐ Define mutualism, commensalism, and parasitism and give examples of each.

☐ Explain "cleaning symbiosis."

☐ Describe a food web and give an example.

Key Terms and Phrases

Symbiosis (Plural is *symbioses*) An intimate association between two or more organisms, such as in mutualism, commensalism, or parasitism.

Mutualism A symbiotic interaction between organisms, beneficial to both parties.

Commensalism A symbiotic interaction in which only one member benefits and neither organism is harmed.

Parasitism A symbiotic relationship in which an organism spends part or all of its life cycle on or within its host and uses the host (or food products within the host) for food.

Free-living Nonparasitic.

Species-specific Implying an exclusive relationship between two species. Parasites are usually species-specific, indicating that they can parasitize only one species of host.

Food web A construct used to analyze feeding relationships within a community. If you are using the Garrison text, look at the examples of food webs on pages 350–351. If you are using the Ingmanson & Wallace text, see the example of a food web on page 349, Figure 16.2.

Biomass A collective term for the living mass of organisms within a specific area.

Mimicry A condition in which two kinds of organisms bear a deceptive resemblance to each other. This resemblance provides a survival advantage for one of the species involved in the relationship.

Before Viewing

☐ If you are using the Garrison text, read "Symbiotic Interactions and Dependencies" on pages 433–435.

☐ Watch Program 20: "Living Together."

After Viewing

Reread the text sections, being sure you understand the key terms and phrases, and complete the following exercises:

1. Is mutualism a valid concept? Are there any examples of truly equal mutualism?

2. Does a host/parasite relationship ever result in invariable death to the host?

3. Can species-specificity rules ever be broken?

4. Regarding cleaner wrasses and mimicking blennies, doesn't the larger fish being cleaned ever become wary of the whole cleaning routine? Might not the larger fish fight back and eat even the cleaners? Does this occur?

Answer Key for Exercises

1. Probably mutualism is a valid concept. In most observed cases of mutualistic relationships, one organism seems to have the better deal, but neither organism lacks benefit from the association.

2. Yes. This can occur when the wrong species of host is infected by a parasite (see next question). It can also occur by design. An example of this is the rabies virus that invariably kills the host after it becomes fully established within the host organism. By the time of the host's death, however, the life cycle of the virus is complete and it has moved to another host, one step ahead of disaster.

3. Yes, often with catastrophic results. As an example, consider the case of the lung flukes that inhabit the respiratory system of most sea lions. If the sea lion is weakened or near death, the feedback system that the parasites use to appraise the strength of the host will

inform them of the fact, and they begin preparations to abandon ship. A sea lion in this condition usually selects a beach on which to rest, and if your pet dog should discover the animal there and sniff at its nose, some of the parasite might make the transfer from sea lion to dog. They quickly travel to the lungs of the dog and set up shop. But the dog is not the specific host for the lung parasite. The biochemical machinery that tells the parasite that it is weakening its host too much is missing. The dog will die a painful death in a few weeks from an overinfestation of lung flukes. The parasite will, of course, also die. The interaction would be a failure in all respects.

4. Not too often. There are relatively few blennies and a relatively large number of wrasses. Perhaps the larger fish consider an occasional fin nip or gill gash a minor price to pay for the alleviation of the irritation of surface parasites. If the relative balance of cleaners and mimics were to change, however, and many more mimics came to the cleaning stations, we could expect the situation to become unstable and the whole cleaning business to become less profitable for all concerned.

Optional Activity

Visit an aquarium, oceanarium, or marine aquarium store and observe the clownfish/anemone association.

Self-Test

1. Which of the following statements is true of symbiotic relationships?
   a. They are unusual and infrequently observed in nature.
   b. They are only occasionally observed in the marine environment.
   c. They are present in marine biology but conspicuously missing in terrestrial biology.
   d. They are the most frequently observed lifestyle of the earth's animals.
   e. They are the most frequently observed lifestyle of the earth's animals because all organisms of the earth live in direct symbiosis.

2. The type of symbiosis implied by food webs is
   a. dependent
   b. direct
   c. facultative
   d. indirect
   e. independent

3. The mode of symbiosis represented by the clownfish/anemone relationship is
   a. parasitism
   b. commensalism
   c. acclimatization
   d. relationism
   e. mutualism

4. The mode of symbiotic relationship represented by the remora/shark association is
   a. commensalism
   b. mutualism
   c. cannibalism
   d. parasitism
   e. catastrophism

5. Intestinal nematode worms within a fish would most probably be an example of
   a. mutualism
   b. parasitism
   c. commensalism
   d. mimicry
   e. antidisestablishmentarianism

6. Species-specific means
   a. a specific host can be parasitized by only one species of parasite
   b. a specific parasite can parasitize only one species of host
   c. only one individual parasite can grow within the host organism
   d. the host can never break free of the parasite
   e. none of the above

7. Which of the following describes the categories of direct symbioses?
   a. They are clearly defined.
   b. They are rarely observed.
   c. They blend somewhat – for example, mutualism can be slightly commensalistic, and so on.
   d. They are impossible to separate.
   e. none of the above

8. Indirect symbioses are primarily involved with
   a. parasites
   b. commensals
   c. benthic organisms
   d. mutuals
   e. food webs

9. The organism that lives within the cells of coral organisms and is capable of forming coral reefs is a (an)
   a. alga
   b. species of worm
   c. type of wrasse
   d. fungus
   e. tiny clam

10. Which of the following statements concerning cleaning symbiosis is *not* true?
    a. Wrasse are the cleaners; blennies are the mimics.
    b. Wrasse greatly outnumber blennies.
    c. Blennies cause grave damage to the larger victimized fish.
    d. The mimic is capable of changing color and behavior to fit the situation.
    e. Cleaning associations are generally tropical or temperate phenomena.

Supplemental Reading

Limbaugh, C., "Cleaning Symbiosis." *Scientific American* (1961).

Childress, J. J., H. Felbeck, and G. Somero, "Symbiosis in the Deep Sea." *Scientific American*, May 1987, 115–120. (The high density of life near hydrothermal vents is explained by mutually beneficial symbioses.)

Wickler, W. *Mimicry*. Translated by R. D. Martin. New York: McGraw-Hill World University Press, 1968.

Lesson Twenty-one   Light in the Sea

Overview

Earth is the blue planet, the sapphire of the skies. It is Neptune's realm – the planet of the sea – and the blue color is just one facet of the remarkable interplay between light and the sea.

Long ago sea people learned to read the mood of the oceans from the display of color: leaden gray under lowering skies; emerald green teeming with life; cerulean blue in warm tropic waters. Today when coastal waters turn a strange red brown, sea people know the red tides have come in. The fish will die and the succulent clams and oysters become lethally dangerous.

This ability of water to reflect, refract, and transmit light brings life to this planet. Every breath we take includes molecules of oxygen released by microscopic marine plants during photosynthesis, using solar light transmitted through the transparent waters. However, water is not perfectly transparent. If it were, the sun's rays would illuminate the greatest depths of the sea, and forests of plants would fill the ocean basins.

The zone of light penetration is called the euphotic zone. In the open sea this may be no more than the upper 100 or 200 m (330–660 ft); in the turbid coastal waters, less than 20 m would be usual. Primary production of food and oxygen for all the creatures that inhabit the sea is limited to these upper waters. What a thin skin this is if we consider that the average depth of the ocean is about 4,000 m (13,120 ft), with the deepest parts extending to over 11,000 m (over 36,000 ft). Most of the world ocean is in the aphotic zone where the light intensity is less than one percent of entering light.

The amount of light penetrating the sea is influenced by a variety of conditions. The solar radiation actually reaching the sea surface depends on the angle of the sun, atmospheric dust, cloud cover, smog, and gases, all of which absorb, reflect, and scatter a portion of the incoming light. In the winter at high latitudes the angle of the sun is so low that the sea surface reflects most of the rays back into space.

Below the sea surface penetration is influenced by substances dissolved in the water, by suspended sediments, and by plankton populations. The red tides, for example, are blooms of dinoflagellates in the surface waters. Not only do the organisms impart the red color to the water, but they also shade the lower layers, preventing some of their own and other species of plankton from photosynthesizing. The murky brown waters seen after storms are colored by sediments stirred up from the bottom or washed in from flooding streams. The deepest penetration of light is in the clear blue waters of the tropics which are relatively free of sediment, plankton, and even dissolved nutrients.

The sunlight that arrives at the sea surface includes all colors of the visible spectrum from red to violet. Each color is determined by a specific range of wavelengths, the shortest waves appearing violet, the longest red. Seawater differentially absorbs the various wavelengths; the ends of the spectrum, both violet and red, are absorbed first. Less than one percent of these colors can penetrate the sea beyond a few meters. Clear seawater is most transparent to the blue and green portions of the spectrum, and divers and underwater photographers are familiar with the bluish cast of the waters. These are the only wavelengths that penetrate the lower euphotic zone.

This differential penetration of light has some interesting effects. In the well-lit surface waters, many fish are either silvery white or exhibit countershading with dark blue-green on the dorsal surface and silver or white on the underside. To a predator looking down, the dark surface of the back is essentially invisible against the dark waters below. A predator looking up will not see the silvery underside in the sunlit waters above. Many fishes and invertebrates living in the lower euphotic zone are red: rockfish, shrimps, sea urchins, sea cucumbers, sea stars, and others. You might think that their bright color would make them more visible to their predators, like a red flag before a charging bull, but it doesn't. Red light does not penetrate more than a few meters; therefore, no red color is reflected off their bodies. They are no more visible as red creatures than they would be if they were green, blue, or even black!

Another point to consider is that the chlorophyll in plants responsible for photosynthesis absorbs and utilizes light energy from the violet and red ends of the spectrum and reflects back the green wavelengths that are not used, hence the green color. In the sea, however, red and violet are quickly absorbed before the energy can reach the plant cells. Chlorophyll is thus much less effective in the sea than on the land. As a result the evolutionary response of most marine plants has been to develop a variety of accessory pigments that function as light-gathering antennae to absorb light energy over a wide range of wavelengths. This has led to a zonation of marine algae by depth.

The green algae, using the red wavelengths, are limited to the upper few meters and can be seen growing on rocks and pilings. The brown algae, including the giant kelps, utilize yellow, green, and blue light, and survive well from the intertidal zone to depths of 50 m. The red algae, which utilize the low intensity, deeply penetrating blue light, have the greatest range and can be found from the rocks on the beach to depths as great as 150 m in clear water.

Light in the sea provides energy for the production of food and oxygen. Changes in available light also seem to trigger such animal responses as feeding, breeding cycles, metabolic rates, growth rates, migration (including daily vertical migration of many fishes and invertebrates), and other behavioral traits. In the deeper sea, strategies for reproduction, locomotion, predation, and predation avoidance (including the ability to produce living light or bioluminescence) seem to be related to the presence or absence of light.

One of the most intriguing adaptations of organisms to living in the dark waters is bioluminescence, or the production of light by living organisms. The enchanting firefly, one of a very

small number of terrestrial bioluminescent species, was celebrated by Chinese poets over 3,000 years ago. The earliest deep-sea explorers in the 1920s and 1930s were delighted by the flashes of light produced by most of the animals living below about 300 m. In an environment dominated by darkness, great pressure, and near-freezing temperatures, bioluminescence plays a crucial role in the survival of the species. Among the lower forms of life in the sea, bioluminescence occurs in bacteria and numerous planktonic forms. It is also found in about half of the animal phyla. Fishes are the only vertebrates to have this ability.

In producing bioluminescence, chemical energy is converted into light energy by the organisms without loss of heat; hence the name "cold light." In some organisms, such as the firefly, the process is almost 100 percent efficient. This is compared to the incandescent light bulb, which is 10 percent efficient with 90 percent of the energy wasted as heat.

Among the organisms producing light, some deep-sea fish, with unusual and highly specialized light organs, have a truly bizarre appearance. The angler fish is an interesting example. It carries its lighted lure on the end of what looks like a fishing pole fastened to its head. The lure conveniently dangles near its mouth. Other fishes that resemble nonluminous species may suddenly emit light from luminous organs within their bodies. A translucent muscle tissue in their sides acts as a lens.

A very common fish on the Pacific coast is the midshipman (*Porichthys*), so named for the spectacular array of more than 700 light-emitting organs, or photophores, on its underside. The photophores are arranged like the ornamental brass buttons on the blazers of midshipmen of the British Navy. The light-emitting capacity of these buttonlike photophores is apparently under the animal's hormonal and neural control. Other fishes, like the little lanternfish, have light organs arranged like the portholes along the sides of a ship. The patterns of light formed by the photophores vary from species to species. The spots on the sides and underside of the lanternfish emit a soft glow, just enough to match the dim light penetrating from the sea surface. The lanternfish does not cast a shadow and is less likely to be seen by the watchful predator below.

One of the most unusual methods of emitting light has been discovered in the flashlight fish. These fish have special pouches under their eyes filled with luminescent bacteria, which the fish cares for and nurtures. Each individual fish raises its own clone of bacteria. There is also a lid under the eye that the fish can raise and lower to turn the light off and on. The beam is relatively powerful, hence the name of the fish.

It is interesting to contemplate the adaptive values of bioluminescence in each of these different animals. The angler fish is surely attempting to attract prey with its shaking, gleaming lure. The specific patterns of porthole lights on the sides of the lanternfish are probably for species identification in finding a mate. The squids of the deep sea emit a glowing cloud of luminescent material to confuse predators or turn away a light-avoiding species. The flashlight fish may just want to light up its surroundings. The midshipman is a puzzle, however, because the fish lives in the euphotic zone and can readily see its prey, its predators, and its own species. Some oceanographers believe the photophores of *Porichthys* are used during courtship as part of a ritual involving light production, biting, and "humming." Perhaps the shining buttons are a relic of another era when the midshipman inhabited deeper waters.

There are also many luminescent plankters, tiny creatures that emit a tiny sparkle when disturbed. These are the organisms that impart the glowing color to breaking waves, to the wake of a ship, or to your footprints in the wet sand at night. We wonder why bioluminescence developed in these microscopic creatures. The adaptive values are certainly far from obvious.

The interplay of light from the sun in the upper waters of the sea and the living light in the dark waters below continues to fascinate poets and scientists alike. Even the rainbow, the ultimate spectacle of color on earth, is made of waters evaporated from the seas and illuminated by the sun, just another manifestation of the complex and intriguing phenomenon of light in the sea.

Learning Objectives

After completing the reading assignment and viewing the program, the student will be able to:

☐ Recognize the physical factors that influence the penetration of light in the sea.

☐ Describe the relationships between plants and light in the sea.

☐ Indicate the interactions between animals and light in the sea.

☐ Discuss bioluminescence and its adaptive values, knowing some of the marine organisms that exhibit this ability.

Key Terms and Phrases

Euphotic zone The upper waters of the sea, varying in depth from about 20 m to about 200 m, in which there is sufficient sunlight for photosynthesis to take place. This is the zone of primary food and oxygen production. There is no euphotic zone in polar seas in the winter because there is no sunlight then. The deepest euphotic zone is in the tropics where the water is clearest and the sun is at the highest angle to the sea surface.

Aphotic zone A zone in which there are no plants because the light is insufficient for photosynthesis to take place. The light intensities in the aphotic zone are so low that only animals with special visual adaptation can detect any rays at all. It also includes areas of total darkness. Most of the ocean is included in this category.

Compensation depth A zone just above the lower limit of the euphotic zone where the rate of photosynthesis is balanced by plant respiration. There is no net gain in food or oxygen production at this depth. The depth varies depending not only on sea and sky conditions, but also on the different species of phytoplankton, each of which has its own peculiar compensation depth.

Algae Nonvascular plants lacking true roots, stems, and leaves. Algae do not produce flowers, fruits, or seeds. They do contain chlorophyll and photosynthesize. Most plants of the sea, including the giant kelp and other seaweeds, are algae.

Bioluminescence The emission of light by living things. The firefly is the best known example of bioluminescence, although very few other terrestrial species have this ability. Bioluminescence is very widespread among animals in the sea. The chemical requirements for light emission may be similar among many very different organisms. Light is usually produced by chemical oxidation in which an organic compound, luciferin, reacts with oxygen in the presence of an enzyme, luciferase. The products of this oxidation reaction are light, carbon dioxide, water, and oxyluciferin. The light produced is in the part of the spectrum visible to the human eye. Little of the energy is released as heat.

Before Viewing

References for this program are scattered through both texts, but pertinent material may be found as follows:

☐ In Garrison, read the information on refraction on pages 158–159, and continue with "Light in the Ocean" on pages 159–161. Note Figure 6.18 showing the progress of light through the ocean. Other references to light appear on pages 166, 334, 349–355, 362–365. In Ingmanson & Wallace, read "Light and Color" in Chapter 8, pages 116–119. Note Table 8.3 showing penetration of light through the sea surface. Other references to light appear on pages 334–335, 339–341, and 350–361.

☐ Watch Program 21: "Light in the Sea."

After Viewing

Reread the Overview and make sure you know the key terms and phrases of this lesson. Reread the text references, looking up any words you don't understand in the Glossary, and complete the following exercises:

1. You are in a submersible to observe a coral reef in the tropics. Without additional light, how far can you descend before it is too dark to discern features of the reef? What factors will determine the limits of light penetration in the sea? Why does the water here seem so blue compared to coastal waters off the mainland? Will you see the giant kelp living here? What might affect their distribution? Will you see green algae growing on the deeper parts of the reef? Explain the distribution of marine plants in depth.

2. Many animals recovered in deep trawls are red. How can they resist heavy predation with such a vivid color?

3. Why is bioluminescence called "cold light"? Why is the firefly more efficient than a light bulb?

4. Compare the various methods of producing bioluminescence; how is light produced by the flashlight fish as compared with the midshipman? What are some of the adaptive values of producing light in the sea?

5. What properties of water make it difficult to see through the air-water interface? Why is it difficult to focus a camera underwater?

6. Many underwater photographs look quite blue, but some show brilliant color. How do you account for each condition?

Optional Activities

1. Check such magazines as *Skin Diver*, *Sea Frontiers*, and others that feature underwater photographs. Try to identify zonation of plants by color and responses of animals to low light intensities. Look at the eyes of the fishes. Are they extra large, reduced in size, placed in an unusual position on the head, or lacking entirely? Actually, all of these conditions do occur.

2. Look at pictures of some small animals of the intermediate depths, below or in the lower euphotic zone. Many are clear, jellylike, or transparent. What is the adaptive value of such a common trait?

3. If you live near a beach, walk in the wet sand some night, especially during a red tide. Your footsteps will glow, as will the wake of a passing ship and the foam of the breaking waves. What causes this lovely sight?

Self-Test

1. Which of the following statements applies to production of food and oxygen in the ocean?
   a. It is abundant, top to bottom.
   b. It is abundant in the surface, but adequate at all depths.
   c. It is limited to the upper 200 m of the sea.
   d. It is limited to the deep water below 200 m.
   e. It is limited to polar seas in summer.

2. The oxygen you are inhaling at this moment
   a. was most likely produced by some diatom in the sea
   b. may be millions of years old
   c. was not around when the earth was first formed
   d. might have been inhaled by Caesar, George Washington, Einstein, and a host of others
   e. all of these

3. Most of the world ocean has the properties of the
   a. euphotic zone
   b. tropical surface waters
   c. coastal water
   d. subpolar open ocean
   e. aphotic zone

4. The depth to which light in the sea can penetrate depends on
   a. the dust, cloud cover, and gases in the atmosphere
   b. the angle of the sun above the horizon
   c. conditions of the sea surface
   d. suspended sediment or plankton in the water
   e. all of these

5. The wavelengths of light that penetrate deepest into the sea are
   a. red and violet
   b. yellow
   c. brown
   d. green and blue
   e. all wavelengths

6. The algae most commonly found on the rocks in shallow water zones are
   a. red
   b. green
   c. brown
   d. lavender and purple
   e. none of these

7. Which statement is true of the compensation depth?
   a. Green algae are the dominant plants.
   b. Photosynthesis does not occur due to lack of light.
   c. Respiration does not occur due to lack of oxygen.
   d. The rate of photosynthesis balances the rate of respiration.
   e. Kelp cutters are compensated for harvesting *Macrocystis*.

8. The red fishes and invertebrates in the deeper euphotic zone
   a. appear red to frighten away predators
   b. are actually endangered species because of heavy predation
   c. are actually blue but, like cooked lobsters, turn red due to increased water temperature
   d. are actually green but change to red when brought up in the net
   e. are actually invisible at these depths because red light does not penetrate this deeply

9. Which of the following accurately describes bioluminescence in the marine environment?
   a. It is a widespread characteristic of many deep-sea animals.
   b. It is rare, but is very common in terrestrial animals.
   c. It is a dominant trait of animals in the euphotic zone.
   d. It is heat-producing and modifies the temperature of the deep sea.
   e. It is uncommon among fish, but common in other groups of vertebrate animals.

10. The flashlight fish produce their light in which way?
   a. through luminescent plankton they feed on
   b. through chemical action in light organs embedded in their skin
   c. by growing luminescent bacteria in special pouches under their eyes
   d. by heating certain chemicals in pouches under their eyes until they glow
   e. by holding *Noctiluca* in front of them like a flashlight

11. The midshipman is unusual because
   a. it is the deepest fish to show bioluminescence
   b. it glows all over like a light bulb
   c. much of its light is lost to heat
   d. it is luminous and lives in shallow, well-lighted waters
   e. it is the smallest organism to exhibit bioluminescence

Lesson Twenty-two   Sound in the Sea

Overview

Long before Ulysses was beguiled by the songs of the Sirens, seafaring people were fascinated by mysterious sounds emanating from the sea. It was Aristotle, almost 2,500 years ago, who first noted that sound was carried by water as well as by air. Leonardo da Vinci, anticipating modern underwater communication, commented that the sound of distant ships could be heard if a long tube was held to one's ear with the other end in the water.

In the late eighteenth and early nineteenth centuries, a number of ingenious and undoubtedly hilarious experiments were carried out to measure the speed of sound in water. The sound of bells, gunpowder, hunting horns, shouts, whistles, alarm clocks, and pistol shots rushed toward underwater receivers – the ears of the experimenters.

An early practical use of sound waves in the sea occurred with the installation of submerged bells on lightships. The noise from these bells, detected through a stethoscope mounted in the hull of a ship, gave warning of danger ahead.

The need for antisubmarine detection in World War I mobilized the scientific resources of the United States to develop new and improved acoustical and electrical devices for use underwater. Today, the best source of information about the sea and its inaccessible bottom terrain is from instruments that utilize some form of sound waves. Sound waves are truly the "eyes" of the oceanographer.

The speed of sound in water is as much a matter of concern to modern researchers as it was to early whistle-blowing and bell-ringing amateur scientists. Sound travels about 334 m/sec in air, but averages a much faster 1,445 m/sec in the sea (and still faster in solids, about 5,104 m/sec in aluminum and 6,000 m/sec in granite). However, the velocity of sound in the sea is variable, increasing with elevations in temperature, pressure, and salinity. In the upper mixed layer, the speed of sound will be near its maximum of about 1,550 m/sec, the result of high water temperatures. In the thermocline below, extending to depths of about 1,000 m, the rapid drop in temperature causes the sound waves to slow until they reach a minimum velocity near the bottom of the thermocline. Below this depth the temperature is uniformly low, but the pressure continues to increase, and thus there is a corresponding increase in the speed of sound.

Sonar (*so*und *na*vigation and *r*anging) used to map the seafloor depends on very accurate measurements of the speed of sound at all depths. A sound pulse or "ping" is sent from a ship to the seafloor and is reflected back to a receiver on the ship's hull. The time it takes for the ping to be transmitted and the echo to be received is recorded. By knowing this travel time plus the exact speed of sound, we can calculate the distance to the seafloor. The doughty lads on HMS *Challenger* would have spent half a day securing this information with a sounding line and lead weight. Now it takes seconds using sound.

For sonar mapping of the seafloor or for determining the distance to objects in the water, a relatively weak high-frequency sound pulse is used. But to determine the thickness, orientation, and nature of the rocks and sediments under the seafloor, strong low-frequency signals needed. Shock waves produced by explosions or seismic air guns travel through the water and penetrate the layers of the sub-bottom. Sound waves move at different speeds through unconsolidated sediments, sandstones and limestones, and crustal basalt. The pulses reflect off the boundaries between the layers of rock and return to the receiver. From these seismic profiles, buried salt domes, ancient coral reef structures, faults, folds in deep sedimentary rocks, and other hidden features of the ocean floor are revealed to the marine geologist.

Certain characteristics of the way sound travels in water limit or enhance its applications for various tasks. For example, if a pinger is lowered into the deep sea and set off, the sound waves will travel to the sea surface but will not penetrate the air-sea interface. The waves will be reflected to the seafloor and bounce back, making several round trips before being dissipated. If all the sounds generated in the sea *could* pass through the sea surface, what a cacophony would fill our ears!

One problem with sonar mapping is that the energy of the sound waves decreases rapidly away from the source. This "propagation loss" is caused by absorption of energy by water molecules and the scattering and spreading of the sound waves.

However, if sound waves are generated in the minimum velocity zone near the bottom of the thermocline, they tend to bounce off the layers of water above and below that have different densities. In this sound-concentrating band, the sound waves can travel great distances horizontally with a minimum loss of energy, offering interesting possibilities for long-distance communication. In a test of this deep sound channel, or SOFAR (Sound Fixing and Ranging) channel, a depth charge set off near Australia was recorded near Bermuda, halfway around the earth!

In addition to detailed mapping of the seafloor, acoustical methods are used today to study currents, eddies, internal waves, and many other physical phenomena of the sea. Underwater sound has many commercial applications and is used extensively in exploration for minerals and oil-producing structures. Even fish are being located by sonic detectors, a very promising new development.

Another aspect to the study of sound in the sea takes us back to Ulysses and the songs of the Sirens: animal sounds in the sea. Some eminent researchers, such as Roger Payne and Sylvia Earle, have suggested that the Siren's songs were possibly whale songs, amplified by the wooden hulls of the ancient Greek vessels. Was it the haunting music of the humpbacked whales living at that time in the Mediterranean that delighted our Greek wayfarer?

Whale sounds were first identified by whalers who sometimes heard strange grating noises through the bottom of the boats. Study of whale sounds using underwater acoustics began in the 1950s. Today there are hundreds of hours of recorded sound from whales and other marine animals.

The actual methods used by whales to produce sound and the ways in which echoes are received and processed are still not fully understood. We know that toothed whales apparently do not have vocal chords, and in some whales sound production seems to come from a site in the forehead. It also is not certain if all whales use echolocation.

One of the most fascinating sources of sound in the sea is the great sperm whale, *Physeter catodon*, that inspired *Moby Dick* and many other tales of adventure. The sperm whale is a toothed whale, the largest of the group Odontoceti, reaching a length of 18 m (60 ft) and weighing as much as 50 tons. They are the deep divers among the whales. One sperm whale, watched by sonar, was observed to go below 2,250 m (7,380 ft). In another case, a bottom-dwelling shark was found in the stomach of a sperm whale that was captured in an area where the seafloor depth was greater than 3,000 m (9,840 ft). Although sperm whales can stay submerged for 45 minutes or more, they must surface for air because they are mammals. They feed principally on the giant squid that inhabit the deep waters, locating their prey in the inky blackness by echo and stunning or even killing them by blasts of intense sound (see Lesson 19).

Sperm whales are not singers, however, and do not produce the squeals, whistles, moans, or melodies of their cousins, the humpbacks. They produce click patterns or codas that are as individual as signatures and can apparently be heard over several kilometers. The conversation between two sperm whales has been recorded, and while not particularly scintillating, it is interpreted to communicate such information as size, speed, direction of movement, how close, how deep, and of course, who.

We have learned much about the behavior of the sperm whale through the use of sound. We know they are aware of us because they respond to the sounds of a sonar pinger by becoming silent. Further studies are continuing with a certain sense of urgency, for the magnificent sperm whale is on the endangered species list, having been hunted almost to extinction.

The humpback whale, a baleen whale, produces what has been called nature's longest, and perhaps slowest song, replete with rumbling low-frequency passages, high-pitched squeaks and squeals, a few doleful groans and gurgles, and some phrases of great beauty. Whales can remember their songs from year to year, although they compose new themes and variations each season.

The exact purpose of the humpback's elaborate songs still eludes scientists. They may be love songs, because they are sung only by adults during what is apparently the breeding season. Apparently all toothed whales, some baleen whales (such as grays, blue, minke, and humpback), and some pinnipeds (such as the Weddell seal and possibly the walrus) do use sound to find food, locate the seafloor, communicate socially, and generally evaluate the nonvisible portion of their surroundings.

It is also interesting to think about how and when whales discovered the deep sound channel to keep in touch with each other over vast distances. With further research, perhaps some day we will be able to sing to these wonderful creatures, using our pingers to make whale-talk. Then what tales will be exchanged by these new Ulysses, human and cetacean, through the medium of sound in the sea!

Learning Objectives

After completing the reading assignment and viewing the program, the student will be able to:

☐ List the physical factors that influence the velocity of sound in the sea, and tell the effects of each.

☐ Explain active and passive sonar and tell how each is used.

☐ Describe the deep scattering layer.

☐ Discuss the causes and importance of the shadow zone and the deep sound channel.

☐ Compare and contrast the sound produced by the sperm whales and the humpback whales and tell how the sounds are used.

Key Terms and Phrases

Sound waves Sound waves are small pressure changes in solids, gases, and liquids. Sound, unlike light, must be carried through a medium.

Sound frequencies Sound waves not only travel at different velocities, but also are produced at different frequencies, or cycles, per second. One cycle per second is called a hertz, after a well-known nineteenth-century physicist. High-frequency sounds, those waves with small wavelengths, are greatly weakened by seawater and can be transmitted only a few tens of meters. Low-frequency waves can be transmitted great distances. A tone of 440 Hz, which would be middle A on a piano, can be transmitted hundreds of kilometers in the sea.

Passive sonar A system for listening to sounds in the sea, either noises made by humans or noises from organisms. It can be used to determine presence and relative direction to sound source.

Active sonar A system in which a sound wave or ping is emitted and the reflected echo recorded. This is the principal method of locating objects underwater, as well as recording continuous profiles of the seafloor.

Shadow zone Sound waves are bent or refracted as they pass through layers of different density in the sea. Just below the mixed layer, at depths of about 80 m, there may be a layer in which no sound waves penetrate. They are bent upward or downward by differences in density. Submarines can avoid detection by sonar in this acoustical shadow.

SOFAR channel or deep sound channel Acronym for sound fixing and ranging channel, a layer within the lower part of the thermocline at depths of about 1,200 m in the North Atlantic and about 600 m in the Northeast Pacific. This is a zone of minimum sound velocity in which sound waves generated within these depths reflect back into the layer rather than penetrate the surrounding waters of different density. It provides a sound-concentrating channel through which sound may be transmitted thousands of kilometers.

Deep scattering layer (DSL) Layers within every ocean that return diffuse scattered echoes, sometimes interfering with sonar mapping of the seafloor. Research during the 1940s and 1950s showed that the echoes were being returned from marine organisms that contained gas bubbles, air bladders, or droplets of oil. These organisms occur in horizontal layers that migrate downward during the day to about 400 m, then return to near surface waters at twilight and early evening. Visual observations from submersibles revealed that several kinds of fish and many kinds of invertebrates are involved in these daily cycles of vertical migration. Variations in light intensity apparently stimulate the migrations. A full moon can cause the DSL to descend; on cloudy days it will rise.

Echolocation A method used by certain marine mammals to locate objects in the sea through the use of sound waves. Sound is emitted and passes without interruption through the water until it encounters a solid object. The waves are then reflected and return to the source. Continual production of clicks or other sounds and evaluation of the reflected waves during swimming give the animal a constant check on all objects in its surroundings. Both low frequency and high frequency waves are produced, giving great detail about the lightless environment. The very large brains of the toothed whales, for example, are probably necessary for rapidly processing the received acoustical information.

(Remember, marine mammals are all air breathers, but many feed upon squid and other deep-sea creatures. They must be able to quickly locate their prey in the darkness, capture it, and return to the surface to breathe.)

Before Viewing

☐ In Garrison, read "Sound in the Ocean," pages 161–165. See also the information on sonar on pages 164–165, on the deep scattering layer on pages 428–435, and reread about the use of sound by cetaceans on pages 403–407. Note also Box 15.7, page 407. In Ingmanson & Wallace, reread "Depth Recording" on pages 36–38. In Chapter 6, note Figure 6.12. How was the information obtained to construct these diagrams? Read "Sound in the Sea" on pages 119–122, reread "Cetaceans" on pages 325–329, and read about the deep scattering layer on pages 356–357.

☐ You may also wish to review Lesson 19 in this study guide.

☐ Check the Glossary for the following words:
deep scattering layer   sonar
echo sounding   sonic marine animals
pinger   sound
seismic profile   sounding
shadow zone   sound velocity
SOFAR

☐ Watch Program 22: "Sound in the Sea."

After Viewing

Reread the text, being sure you understand the key terms and phrases. The diagrams in this section are very helpful. Complete the following exercises:

1. Why is the speed of sound in the sea of prime concern to the oceanographer? What are the main factors that affect the speed of sound and, in each case, how do they alter the velocity?

2. How could you tell if a sound in the sea was from an organism or from breaking waves? If you are using Ingmanson & Wallace, see Figure 8.15 on page 121. Check *sound frequencies* under key terms.

3. Why is the shadow zone of great interest to the navy?

4. What might be some practical applications of the SOFAR channel?

5. When two sperm whales approach each other in the blackness of the deep sea, they exchange click sequences called codas. What kinds of information might they be exchanging?

6. In addition to marine mammals, what other animals produce sounds in the sea? Do any land animals echolocate?

Optional Activities

1. Listen to records of whale songs. A recording is included in the January 1979 *National Geographic*, along with articles by Sylvia Earle and Roger Payne. Also see the listing

of recordings in Lesson 19 of this guide. Can you detect repeated phrases and sequences?

2. The next time you are in a pool, have a friend click two stones together while you are submerged. You will be rediscovering what Aristotle knew 2,500 years ago.

3. Can you think of a method that eighteenth-century amateur scientists might have used, other than those listed in the Overview, to accurately measure the speed of sound in water?

4. Read the collection of articles on "Sound in the Sea," in the Spring 1977 issue of *Oceanus*, published by the Woods Hole Oceanographic Institution.

Self-Test

1. Comparing the speed of sound in solids, liquids, and gases, we find it is slowest in
   a. gases
   b. solids
   c. liquids

2. The velocity of sound in the sea increases as the
   a. temperature is lowered
   b. pressure increases
   c. salinity becomes less
   d. surface waters become colder
   e. surface waters become fresher

3. The method used for listening to animals in the sea is
   a. SOFAR
   b. echolocation
   c. precision depth recording
   d. active sonar
   e. passive sonar

4. Before working in the sea, the oceanographer must have a very accurate map of the area. Using the SONAR, he or she must determine the speed of sound in the water and
   a. the water's salinity
   b. the volume of sound
   c. its frequency
   d. the exact travel time of ping and echo
   e. the depth to SOFAR channel

5. The deep scattering layer is so named because
   a. it is made up of scattered animals
   b. it reflects sounds from deep scattered rocks on the seafloor
   c. it scatters reflected sound waves
   d. it produces intermittent clicks
   e. the sharp echo produced causes organisms to scatter

6. In ocean acoustics, the shadow zone is a region
   a. on the deep seafloor where no light can penetrate
   b. within the thermocline where sound reaches its maximum velocity
   c. in which sound waves are very clearly heard
   d. in which sound waves are bent away and little or no sound penetrates
   e. that is a strongly reflecting layer in the seafloor

7. The deep sound channel is characterized as a zone in which
   a. sound is concentrated rather than diffused through the water
   b. energy losses are small
   c. sound waves travel great distances horizontally
   d. sound velocities are at a minimum
   e. all of the above

8. The deep scattering layer can be described by which of the following statements?
   a. It rises at night and descends during the day.
   b. It results from sound waves being reflected from the bodies of many different marine organisms.
   c. It may be one or more layers in the sea.
   d. It is present in most oceans.
   e. all of the above

9. The most musical songs are produced by the
   a. pistol shrimp
   b. sperm whales
   c. bottle-nosed dolphins
   d. humpback whales
   e. white croakers

10. The sounds of the sperm whales are
    a. patterns of clicks
    b. used to locate prey
    c. used to communicate with other whales
    d. used to locate objects in the water
    e. all of these

11. Seismic profiling is used to
    a. send sound waves over great distances in seawater
    b. determine the structure and type of rocks lying below the seafloor
    c. locate schools of fish
    d. map ridges and rises on the seafloor
    e. all of the above

Lesson Twenty-three   Life Under Pressure

Overview

Perhaps no oceanic realm so captures the imagination as the deep-ocean zone. It conjures up visions of huge sea monsters living in its cold depths, of giant squid capable of inflicting severe wounds on the sperm whales that hunt them there, of fantastic glowing fish capable of eating organisms larger than themselves, and of endless sameness. In the deep zones it is eternally dark, forever cold, and always highly pressurized.

The deep waters from about 900 m (3,000 ft) to the ocean floor are quite stable and uniform. The epipelagic zone, the sunlit area containing the abundance of sea life, lies above these deep zones from the ocean's surface to a depth of 150 m (about 500 ft). The deeper water is divided by some oceanographers as follows:

mesopelagic zone   =   150–1,000 m
                                          (500–3,300 ft)

bathypelagic zone   =   900–3,700 m
                                          (3,000–12,000 ft)

abyssal zone   =   below 3,700 m
                                 (12,000 ft)

Some researchers include another special zone within the abyssal zone – the hadal zone, which lies at the bottom of the deepest trenches.

The water in the bathypelagic and abyssal zones is usually very cold (1–2°C, 33–35°F). It is relatively high in salinity (around 36‰), and is lacking in large-scale movement. Although slow currents are perceptible, there is no movement from passing surface waves. The pressure at the bottom is equal to 1,100 atmospheres (nearly 8 tons/in.²). Water in the hadal zone is occasionally slightly warmer than water above because the trenches in which hadal water is contained slash deeply into the earth's crust. The mantle underlying the crust is quite warm (average temperature, approximately 1,800°F), and some of this heat "leaks" through the thin crust into the water near the bottom.

Nearly all organisms living in the deep ocean are dependent on the productivity of the water column above. There is no light capable of supporting photosynthesis in the ocean at these depths, and so no green plants can survive. There are some organisms in special areas of the abyss that live as part of a food web based on bacteria capable of metabolizing the hydrogen sulfide found in water flowing from "geysers" in the seafloor.

Organisms living away from the surface must recycle the rain of organisms, food particles, and organic and inorganic nutrients from above in order to survive. In fact, the feeding adaptations exhibited by animals here are truly astonishing. Gulper eels, for example, have extendable jaws and a stomach capable of consuming prey greater in size than the eels themselves. Tripod fish use very sensitive extensions of their fins and gill covers to detect the vibrations of prey tens of feet away. Some organisms whose mouths blend with the natural contours of the rocky bottom provide living caves into which small bottom dwellers crawl for protection, only to be swallowed alive. Other fish are capable of smelling dead organisms for miles down-current. All must take advantage of the infrequent opportunities for feeding.

Bioluminescence is important in the deep ocean in both feeding and mate attraction. Some abyssal organisms attract meals with a luminous worm-shaped lure. A possible predator comes to investigate and itself becomes lunch! Other animals use a pattern of spots or lines to identify themselves to members of the same species, a necessary first step in mating behavior. Deep-ocean squid even use a luminous jet of "ink" to frighten potential predators. Of course, black ink would be of no use in this dark ocean world.

The physical appearance of organisms from this zone is often strikingly different. Their flesh may be watery and red or cream-colored. They may have long sensory appendages trailing from their bodies, or huge upward-tilted eyes with which to see animals at the very bottom of the sunlit ocean, and long, pointed teeth with which to capture them. Most of the fish are quite light in weight and often require no swim bladders. Because of the greater proportion of dissolved CO_2 at this depth and the consequently acidic pH, the deposition of a substantial skeleton is very difficult. Fish here are quite unlike those found in the upper regions.

The deep zone is not heavily populated. Not many organisms have developed the necessary adaptations to deal with life here. For example, it has been estimated that about 83 percent of the ocean's total biomass (mass of living matter) is concentrated in the uppermost 200 m (660 ft) of ocean and only 0.8 percent is found below 3,000 m (10,000 ft). The deep ocean is sparsely populated, indeed.

Curiously, pressure itself is not the prime difficulty. The organisms have the same amount of hydrostatic pressure within their bodies pushing out as they have outside pushing in. They are no more aware of the pressure at their depth than we are of the 14.7 $lb/in.^2$ of atmospheric pressure beneath which we exist. The cold temperature of the water and lack of nutrients appear to be more critical in limiting the numbers of animals surviving at this depth.

Some of the most exciting and unexpected discoveries in marine biology have been made in the deep ocean. The "*Challenger* Reports," issued by the members of the *Challenger* expedition after its return, repeatedly mentions organisms new to science from most of the 492 soundings made by the expedition. The sense of scientific anticipation when each sampler was emptied must have been tremendous! In 1938 a living coelacanth (genus *Latimeria*) was netted off the east coast of Africa. This large, lobe-fin fish was known only from fossils and was thought to have been extinct for 60 million years. Rewards were offered, and in 1952 a second specimen was obtained near the Comoro Islands. Many research groups are searching for living specimens. One has been photographed recently.

The oceanographic world was shaken in 1977 when scientists inside Woods Hole Oceanographic Institution's submersible *Alvin*, searching at 3,000 m (10,000 ft) for warm water (to 17°C, 63°F) flowing from vents in the young ocean floor crust some 220 mi north and east of the Galápagos Islands, discovered highly developed and dense communities of large animals clustered around the vents. The water was laden with hydrogen sulfide, carbon dioxide, and oxygen upon which local bacteria lived. These bacteria evidently form the base of a food chain that extends to large animals. Unusual wormlike organisms were found in this area. Some of the "worms," measuring 1.5 m (4.5 ft) in length and 5 cm (2 in.) in diameter, were contained in their own long parchmentlike tubes and had a hemoglobin-rich blood.

These strange animals have been identified as pogonophorans, members of a minor phylum of invertebrate animals that also have representatives in shallow water. Three vent species have been identified thus far. The tubes of these Vestimentifera (or "beardworms") are flexible, yet capable of housing the length of the animal when it retracts. Their bodies consist of four regions: tentacles, vestimentium, trunk, and segmented opisthosome. Feeding is something of a puzzle because these animals have no mouth, digestive tract, or anus. The tentacles, however, seem important in feeding. Each worm has about 230,000 tentacles, each containing a pair of blood vessels. The trunk of the worm is probably the key to the nutrition riddle. It is actually a tightly packed mass of bacteria similar to those found near the geothermal vents. Evidently these bacteria, like those living separately near the vents, use hydrogen sulfide as an energy source to convert carbon dioxide into organic molecules. The worm may selectively "crop" bacteria from its own trunk or absorb external bacteria into the outer surface of the worm itself. Either way, the ultimate source of the worm's energy is bacteria, which have, in turn, fed by means of converting the energy in the hydrogen sulfide molecule into usable biological energy (chemosynthesis). The association between worm and bacteria appears to be either symbiotic, or grazer, or both. More research is needed to discover the full nature of the relationship.

In Lesson 2 you learned that some biologists believe life on this planet originated in the shallow seas. Not all researchers share that view. Critics of the shallow seas hypothesis suggest there would have been inadequate concentrations of organic matter available to meet the nutrient needs of primitive cells, the large amount of ultraviolet radiation at the surface might have shattered large molecules, and that the membranes that cells possess would not have formed in that shallow-water environment.

One hypothesis formulated by John Corliss (see the supplemental reading for this lesson) argues that highly concentrated dissolved minerals escaped through holes in the ocean floor that were lined with a chemical catalyst, perhaps a magnesium clay. In the presence of this catalyst, small molecules combined into larger, complex organic molecules, which, in turn, combined to form protocells. He argues that these catalysts and dissolved mineral concentrations were not present at the surface, so life would not have formed there.

Fossils found in rocks formed 3.5 billion years ago tend to support his hypothesis. There is an astonishing similarity between filamentous "organic-like" fossils from Warrawoona in western Australia (see Garrison Figure 1.16, page 15) and organic structures seen in the 1979 electron microscope photographs of hydrothermal deposits taken in the East Pacific Rise. Still, the easy availability of different forms of energy and the stirring action of turbulence in the surface environment lead many to believe biosynthesis did occur at or near the surface.

As you can imagine, to collect organisms and gather environmental data from these depths is quite difficult. Photographs taken of the bottom show that it is covered in some places with brittle stars, and devoid of them in others. Sharks are present in some regions and absent in other seemingly identical spots. Conditions appear uniform, but animal distribution is not. We clearly have much to learn about this unusual oceanic realm.

Learning Objectives

After completing the reading assignment and viewing the program, the student will be able to:

☐ Distinguish among three deep-ocean zones and describe the basic physical water conditions within these zones.

☐ Identify the food sources for deep-water organisms.

☐ Describe the physical appearance of some of the deep-water life forms.

☐ Contrast the deep bottom vent organisms and their food chains to the traditional organisms and their food chains.

☐ Explain why sampling within the deep ocean is so difficult.

☐ Discuss the role of bioluminescence in the deep ocean.

☐ Suggest a feeding method for abyssal pogonophoran worms.

☐ Contrast arguments for the origin of life at the ocean surface with arguments for the origin of life at the abyss.

Key Terms and Phrases

Deep-ocean zones Those areas of ocean into which sunlight never penetrates and in which physical conditions are relatively constant. No primary productivity occurs in the deep zones. (The zone above is known as the epipelagic zone.)

Epipelagic zone The sunlit zone above the mesopelagic zone. Primary productivity (photosynthetic activity) occurs in the epipelagic zone.

Mesopelagic zone That zone of ocean immediately below the epipelagic zone. This zone is the home of many organisms that ascend to the epipelagic zone after nightfall to feed.

Bathypelagic zone That zone of deep ocean below the mesopelagic zone. Its depth is generally given as 900–3,700 m (3,000–12,000 ft). Animals here are usually dark red or black in color and prey on organisms that inhabit the lower limit of the epipelagic zone.

Abyssal zone The water below 3,700 m (12,000 ft). The organisms may be cream-colored and generally light in weight, may have a watery flesh, lack swim bladders, and may eat only rarely.

Hadal zone The deepest ocean realm, that area within deep-ocean trenches.

Bioluminescence Biological light. Among other purposes this cold light is used for mate attraction, food acquisition, and for frightening potential predators.

Deep scattering layer A relatively dense aggregation of fishes and squid and other organisms capable of reflecting a sonar pulse. The reflected pulse resembles a false bottom in the ocean. The scattering layer position varies with the time of day because of the amount of sunlight reaching the lower edge of the epipelagic zone.

Aphotic zone Literally without light. The zone of ocean below the epipelagic zone into which sunlight never penetrates. This term is identical to deep-ocean zones as used in this lesson.

Chemosynthesis The synthesis of organic compounds from inorganic compounds using energy stored in chemicals. This may be contrasted with photosynthesis, the process by which green plants make organic compounds from inorganic compounds by using the energy of sunlight. Since no sunlight is available in the abyss, primary productivity must be accomplished by alternate means.

Before Viewing

☐ In Garrison, read the sections on "Temperature," "Salinity," "Dissolved Nutrients," "Dissolved Gases," "Acid-Base Balance," "Hydrostatic Pressure," and "The Interplay of Factors," on pages 335–337. Note also the distribution of the oceanic provinces on pages 342. Read pages 429–433 discussing "The Deep Sea Floor" and "Vent Communities."

☐ In Ingmanson & Wallace, reread "Pressure" and "Depth," pages 338–340. Also read "Depth, Darkness, Shortage of Food: Life in Deep Water," pages 356–362.

☐ Watch Program 23: "Life Under Pressure."

After Viewing

Reread the text sections, being sure you understand the key terms and phrases, and complete the following exercises.

1. Why are the epipelagic zone and the deep zones so vastly different?

2. Why should organisms from these zones be so different from each other?

3. Do you think that organisms from one zone could, if accidentally introduced into another zone, survive and prosper?

4. Is pressure the major problem in living here? Why do most people list this as the primary concern for these animals?

5. Are there any plants in the abyss?

6. Why was the name "Vestimentifera" first applied to the abyssal pogonophorans?

Answer Key for Exercises

1. The main differences are due to sunlight penetration, nutrient availability, and thermocline depth. Above the bottom of the epipelagic zone the ocean is bathed in light, plants can survive, seasons change, and food is abundant. Below the line, a dark sameness pervades the ocean.

2. Organisms reflect their environment and the conditions of that environment. Different environments yield different organisms.

3. Not likely. They may live for a while, but eventually some condition needed for their survival would prove limiting and they would die.

4. Pressure is not important in terms of being crushed because organisms here have the same pressure inside as outside – there is no pressure differential. Water pressure can affect physiological processes and biochemistry, however. The animals that live in these surroundings have evolved adaptations that permit their proper biochemical functioning under this pressure. Even if a way were found to prevent the collapse of the gas-filled spaces within our bodies, people could not withstand the pressure increase for biochemical reasons. It is possible, incidentally, to slowly increase the atmospheric pressure on a human to many times normal and still allow the person to survive. When people visit the abyssal zones, however, they must go within a thick steel sphere capable of withstanding the pressure difference outside (1,100 atmospheres) and inside (1 atmosphere).

5. No plants can long survive where there is no sunlight. The bioluminescence produced by some of the animals of the deep zones is not enough to allow photosynthesis.

6. Because the discoverers thought the shape and color of the top segment of the worm resembled the ecclesiastical vestments worn by some clergy.

Optional Activity

Museums occasionally have models of deepwater fishes on display. You may wish to visit a museum or aquarium facility and see for yourself what these organisms look like.

Self-Test

1. Water temperature in deep-water zones is best described by which of the following?
   a. is constant at about 10°C (50°F)
   b. varies considerably, but centers around 1°C (34°F)
   c. is constant at about 1°C (34°F)
   d. is constant at about 1°C (34°F), *but* in very deep areas or near vents may be slightly warmer
   e. is below the freezing point of water; however, the water does not actually freeze because the hydrostatic pressure is too high for ice crystals to form

2. Which of the following statements applies to water salinity in deep-water zones?
   a. It is tremendously variable.
   b. It is variable, but centers around 36‰.
   c. It is remarkably constant, centering around 36‰.
   d. It is always lower than that at the surface water directly above.
   e. It is always higher than that of the surface water directly above.

3. Water pressure affects deep-water organisms in which ways?
   a. It squeezes organisms relentlessly; organisms must be tremendously strong to survive in the deep-ocean zone.
   b. It squeezes organisms relentlessly, but makes no difference to them at all.
   c. It squeezes organisms relentlessly, but their physiological adaptations permit them to survive; the actual pressure difference (inside-to-outside) is nil.
   d. It allows humans to enter the truly deep zones with SCUBA-like equipment and return safely.
   e. It prevents any life from living in the extreme depths.

4. Which of these zones is the deepest?
   a. hadal
   b. epipelagic
   c. bathypelagic
   d. neritic
   e. abyssal

5. For which of these purposes is bioluminescence *not* used in the deep zones?
   a. mate attraction
   b. food attraction
   c. warning
   d. frightening off potential predators
   e. photosynthesis

6. The most frequently encountered feeding method in the abyssal benthos (bottom) is
   a. predation
   b. filter feeding
   c. photosynthesis
   d. detritus feeding
   e. cleaning symbiosis

7. Sampling is difficult in the deep zones because
   a. cable lengths are extreme, and the possibility for cable failure is therefore relatively great
   b. positioning the ship over one area of ocean bottom is difficult to accomplish
   c. currents may vary with depth
   d. organisms are infrequently encountered at great depth
   e. all of the above

8. Which of these statements best describes the trophic situation (food) in the deep zones?
   a. Food is plentiful and easily found in the deep zones.
   b. There is virtually no food in the deep zones, and few life forms exist there.
   c. Food is not abundant, but if the waters above are productive, the deep zones can support a surprisingly diverse group of animals.
   d. Food is not abundant, but one or two kinds of animals can usually survive in the deep zones below productive areas.
   e. In some areas of the world ocean there is no food at great depths, and therefore no life.

9. Which of these statements is most nearly correct?
   a. The number and weight of living things (that is, biomass) in the abyssal zone exceeds that in the epipelagic zone.
   b. The biomass in the abyss is about equal to that of the epipelagic zone.
   c. The biomass in the abyss is slightly less than that of the epipelagic zone.
   d. The biomass in the abyss is vastly less than that of the epipelagic zone.
   e. No studies that reliably compare biomass in the epipelagic and abyssal zones have been successfully completed.

10. What do vestimentiferan worms (pogonophoran "beardworms") probably "eat"?
    a. giant squid
    b. diatoms
    c. dissolved organic material
    d. bacteria that receive energy from hydrogen sulfide
    e. bacteria that receive energy from carbon dioxide

Supplemental Reading

Ballard, R. D., "Notes on a Major Oceanographic Find." *Oceanus*, vol. 21, no. 1 (1977), 50–59. (The first published account of the discovery of abyssal vent communities.)

Corliss, J. B., J. A. Baross, and S. E. Hoffman, "An Hypothesis Concerning the Relationship Between Submarine Hot Springs and the Origin of Life on Earth." *Oceanologica Acta* (1981), Geology of Oceans Symposium, Proceedings of the 26th International Geological Congress, 7–17 July 1980, Paris.

Dietz, R. S., "The Sea's Deep Scattering Layers." *Scientific American*, vol. 207, no. 2 (1962), 44–50.

Heezen, B., and C. Hollister, *The Face of the Deep.* Oxford University Press, 1971. (A large-format paperback containing many pictures of deep-sea geological and biological wonders.)

Hollister, C., et al., "The Dynamic Abyss." *Scientific American*, vol. 250, no. 3 (1984), 42.

Isaacs and Schwartzlose, "Active Animals of the Deep Sea Floor." *Scientific American*, vol. 233, no. 4 (1975), 84–91.

Petit, C., "Neptune's Forge." *Science 83*, vol. 4, no. 1 (1983), 60.

Whitehead, J. A., "Giant Ocean Cataracts." *Scientific American*, Feb. 1989, 50–57. ("Waterfalls" of dense water purging down continental slopes are likened to cataracts. The role of dense water in maintaining the chemistry and climate of the deep ocean is explored.)

Lesson Twenty-four   The Polar Seas

Overview

The polar regions of the earth are areas of great contrasts and dramatic extremes. The North Pole pierces the earth's surface through an ocean covered with a light crust of ice; the South Pole through land covered by a layer of ice that averages a mile in thickness. The continent of Antarctica, Earth's fifth largest continental land mass, contain 90 percent of the world's ice, about 7 million cubic miles! In spite of all the ice, strangely enough the earth's polar regions are essentially deserts, experiencing generally low levels of precipitation.

The northern polar regions are dominated by the Arctic Ocean, the southern regions by ice and mountainous terrain. Both polar areas share such unusual illumination phenomena as six months of daylight ("midnight sun"), followed by six months of darkness; auroral displays high in the atmosphere caused by subatomic solar particles spiraling toward the surface within the earth's magnetic field (the "northern lights" and "southern lights"); and magnificent reflection patterns and odd shapes in the air caused by mirage layers and tiny suspended ice crystals.

The mere mention of a polar sea conjures images of icebergs. Arctic icebergs are generally much different in appearance and character from Antarctic icebergs. The former tend to be dense, hard, pinnacle-shaped, and dangerous to shipping (see pages 144–145 in Garrison, or Figure 17.12a, page 383 of Ingmanson & Wallace). Antarctic icebergs are flatter (table-shaped), rather less dense, and float with a smaller percentage of their bulk submerged (see Figure 17.12b of Ingmanson & Wallace). Antarctic icebergs rarely cause shipping disasters.

When the origins of icebergs from the different regions are contrasted, the reasons for the differences become obvious. Arctic icebergs are generated from huge glaciers squeezing down fjord valleys cut into high coastlines. Many large and dangerous bergs form along the southeast coast of Greenland. Greenland is generally higher at its center than at its edges. The Greenland ice cap feeds glaciers that slowly move down toward the coastline. Icebergs "calve" from the glacier ends under the influence of tide and temperature changes. The Pacific area of the Arctic, unlike the Atlantic area, does not possess geographic features that produce icebergs. Also, the Bering Strait limits the flow into the Pacific of what few icebergs are formed. Therefore, it is the Atlantic that receives the brunt of iceberg formation and southern migration.

Some of Antarctica, on the other hand, tends to be higher at the edges than at the center, so ice deposited on the continent often stays where it is rather than move toward the shore. In fact, Antarctica has been greatly depressed, actually pushed into the earth's mantle, by the great weight of this immobile ice cap, which, astonishingly, contains about 80 percent of the earth's entire supply of fresh water. However, tremendous areas of sea surface do freeze around Antarctica's edges, most especially in bays. These frozen bays remain stable for tens of years in some cases, accumulating great masses of ice from the sloping edge of the continent and some of the fallen snow. Eventually, when their weight becomes too great and warm temperatures or storms weaken their edges, they break free from the land and float north and east in the West Wind Drift. The most astonishing example of this long-range tabular iceberg was noted in 1927 when a 10,000-square-mile-chunk of "shelf ice" (as these separated frozen bays are called), about eight times the area of the state of Rhode Island, broke free and drifted north past the Falkland Islands off the coast of Argentina!

The polar oceanic regions are surprisingly productive despite the low temperatures encountered there. Silicates are present in abundance to support the growth of glass diatom shells, and the levels of phosphate "fertilizers" required for healthy phytoplankton are almost always high. Indeed, nutrients of many sorts are plentiful because of the frequent upwelling in these areas. Oxygen is readily available in the polar oceans because the temperature of the water is low, and there is usually considerable turbulence caused by storms in ice-free areas. Of course, phytoplanktonic organisms require sunlight for photosynthesis, and sunlight is available in adequate intensities at these latitudes for only two or three months of the year. On the other hand, during these months the intensity of the bloom is astonishing. The ocean surface for tens of square miles may resemble tomato soup!

The zooplanktonic life in the polar oceans is equally impressive. As an example, consider the fact that one acre of rich terrestrial farmland can support about 700 lb of cattle per acre. In comparison, the polar seas frequently contain as much as 1,000 lb of krill (a species of planktonic arthropod) per acre. This krill is a food source for the largest of whales that populate the polar seas of both hemispheres.

How do the low ocean temperatures affect organisms? Animals and plants living in the polar oceans have longer, slower growing periods and longer life spans than their counterparts in warmer climates. Adults of the same species may be larger in size than similar species found in tropical or temperate oceans, but the number of species actually present in polar waters is smaller in comparison to tropical waters. In general, biological activity is directly proportional to water temperature: the higher the temperature, the greater the movement and metabolism.

Weather near the poles, as one would expect, is quite unstable and often extreme. Perhaps the most turbulent and concentrated frontal storms in the world are those of the West Wind Drift, a deceptively lovely name for the circum-Antarctic ocean area south of the tips of South America and Africa. The almost continuous storms here have frustrated navigators for centuries, especially when they tried to "round the horn" or the Cape of Good Hope. This violent ocean area provided early Antarctic explorers with their first taste of the rigorous adventure to come, and what a taste it could be.

The United States' weather is often influenced by Arctic polar forces. Great fronts of frigid Arctic air sweep out of the region and across the plains states toward the East Coast. If these fronts encounter warmer moist air from the south, resulting winter storms can be devastating.

Interest in polar exploration reached a peak in the early years of the twentieth century with Scandinavian, British, and American explorers probing the mysteries of the ocean and land areas at the top and bottom of the world. Curiously, the breakthrough that allowed successful assaults on the poles and the surrounding ocean areas was not connected with oceanography per se. It involved the perfection of the food canning process. Prior to this development, survival on long treks away from shore (where fresh meat was available) was precarious at best, especially on the Antarctic continent, where sea life is not available underneath an ice pack, as it is in the North Pole area. This advance, plus new ideas in shipbuilding, scientific curiosity, national pride, and personal courage, led to the golden age of polar exploration.

After some genuinely heroic attempts by a number of explorers to reach the poles, an American naval officer, Robert E. Peary, accompanied by his black assistant, Matthew Hensen, and four Eskimo, reached the North Pole in April of 1909. The South Pole was attained in December 1911 by a party of five men led by Roald Amundsen of Norway. In January 1912 the group led by Captain Robert E. Scott of Great Britain reached the same southern goal. The tragic story of Scott's failed attempt to return to base camp remains one of the most dramatic tales in the long history of exploration. In 1958, under the command of Captain Anderson, the United States nuclear submarine *Nautilus* sailed beneath the North Pole in a submerged transit beneath the Arctic ice cap from Point Barrow, Alaska, to the Greenland Sea.

Perhaps the most astonishing saga of polar exploration, and surely the one with the most surprising outcome, is the nearly unbelievable 1914–1916 adventure of Ernest Shackleton (later Sir Ernest), a British merchant navy officer. He had mounted an expedition to try to cross the Antarctic continent from sea to sea. Shackleton's ship *Endurance* was immovably gripped by ice in Antarctica's Weddell Sea on January 18, 1915. Slowly, *Endurance* was crushed by the ice and nine months later she sank. The party of 28 explorers camped on the frozen sea and drifted north for five months until the ice began to break up. They had rescued from the remains of their ship some small boats and supplies which they dragged to open water. After a brief sea trek the party was cast ashore on desolate Elephant Island at the tip of the Palmer Peninsula. This island was so isolated that no hope was held for rescue; someone had to go for help.

Shackleton's largest boat was only 6.85 m (22 ft 6 in.) long, and ahead of him lay the most violent ocean of the earth. Where could they go for help? There was only one chance and that was to attempt a landfall at the whaling station of South Georgia Island over 1,280 km (800 mi) away. With the force of the westerlies there would be no hope of return if they were blown past the island.

A crew of six was selected. No one can imagine that voyage. The incredible discomfort must have been all but unbearable. The skill, fortitude, strength, and courage of these people against such a hostile sea with its mountainous waves is wonderful to consider.

After sixteen days of howling seas they spotted South Georgia Island. Unfortunately, the whaling station was on the east side of the island, and they had made their landfall on the west side. Between them and safety lay a mountain range that had never been traversed. Three incapacitated members waited while Shackleton, Worsley (captain of *Endurance*), and one other man set off across glaciers and snow-choked passes, only to be frequently turned back by impassable terrain and towering cliffs. After nearly three days of agonizing effort, they stumbled into the whaling station at Strømness, only to have the residents reject their story as an utter impossibility and the ravings of madmen! No one could have done what they claimed. The mountains to the east were alone considered an insurmountable barrier. As for this wild story of a voyage across the towering West Wind Drift in a 22-ft boat – that was too unbelievable to be taken seriously.

Eventually the whalers were convinced and a rescue effort was mounted to save those men still marooned on Elephant Island. Winter was approaching and the pack ice was quickly forming. The first three attempts to reach Elephant Island were unsuccessful, but finally, on August 30, 1916, Shackleton broke through and found all twenty-two of his men safe and well. The final note that makes this adventure so marvelous is the fact that *not a single man was lost* and no serious injuries were sustained by any

member of the expedition. By all means read the book by Captain Worsley listed in the supplemental reading for this lesson! Why read fiction when you can read nonfiction like this!

Early polar explorers observed, and ate, large numbers of seals, sea lions, and penguins. Because these animals require easy access to their oceanic food, they are concentrated near the shoreline in the Antarctic and on drifting floes of ice in both regions. Penguins occur naturally only in the southern ocean, and are birds that have through evolution traded the ability to fly in air for the ability to "fly" underwater. They are as spectacularly agile in the water as they are awkward on land. When established in breeding colonies, they exhibit "huddle" behavior (as opposed to marking and maintaining broad territories) to conserve body heat.

Seals in the polar regions grow to considerable size and are experts in deep diving. When seals go in search of food, their dives may last up to 70 min and extend to a depth of nearly 2,000 ft.

Polar bears, a large evolutionary variant of North American bear stock, inhabit the shifting floes and pack ice of the Arctic Ocean. They feed primarily on seals and fish and can be found at the North Pole itself. One of the first things the lookout of the United States nuclear submarine *Skate* saw when the boat surfaced near the North Pole was a polar bear walking nearby.

There are terrestrial animals and plants in high latitudes, but generally very few species. In the entire area above the Arctic Circle there are fewer than 70 species of animals. Forty-four of these are insects, the largest of which is a type of wingless mosquito. There are no land mammals at all in Antarctica, and only five species of seals inhabit the water surrounding the continent. Antarctica does, however, support at least 800 species of plants of which 350 are lichens. Lichens, which are very slow growing, are ideally suited to the rigors of the Antarctic climate. One form of lichen actually grows within rocks!

Antarctica, Earth's most remote and unspoiled continent, provides unique opportunities for scientific investigation. Toward that end, a treaty was signed in 1959 to reserve the entire continent for nonpolitical scientific use. Many important findings have been made there, including the recent discovery of a hole in the atmospheric ozone layer above the continent (see Lesson 28). Remember that 250 million years ago Antarctica was a tropical continent, so it is understandable that great coal deposits have already been found. A consensus of informed opinion holds that oil and natural gas may be present near the edge of the continent in prodigious quantities. The future of this awe-inspiring continent is right now being decided. Let's hope the planners plan wisely.

The polar regions of the earth continue to present a challenge to oceanographers who understand their importance to marine productivity and to meteorologists who see them as the birthplaces of violent frontal storms. To the layperson, the polar oceans are still an unchallengeable frontier that conquers all but the imagination.

Learning Objectives

After completing the reading assignment and viewing the program, the student will be able to:

☐ Recognize the fundamental differences between the Arctic and Antarctic regions and Know the differences between Arctic and Antarctic icebergs.

☐ List three of the unusual illumination phenomena experienced in polar regions.

☐ Explain the causes of high summer productivity of the polar seas.

☐ Understand in general terms how low water temperature affects marine life.

☐ List some large animal inhabitants of the polar region, and note which live in the north, which in the south, and which in both areas.

☐ Outline a brief history of the exploration of the polar regions.

☐ Understand the growing dilemma arising from conflicting needs for the Antarctic continent to be preserved for scientific research, and, on the other hand, the need for minerals and energy, for world distribution.

Key Terms and Phrases

Productivity Productivity in a biological sense implies "first production" of food. The only primary producers of food at the ocean's surface are plants. Oceanic regions of high productivity are therefore areas with abundant marine plants, such as phytoplankton, which form the basis of food webs.

Limiting factor Any physical or biological factor that, when present in an inappropriate amount, limits the ability of a population to survive.

Krill Common name for the euphausiid *Euphausia supurba*, a small (2.5–5 cm or 1–2 in.) planktonic arthropod abundant in polar seas. These organisms form the major food species for the great (baleen) whales.

Arctic A general term for the area north of the Arctic Circle (at latitude 66.5°N) or for the phenomena, objects, or organisms originating in that area.

Antarctic A general term for the area south of the Antarctic Circle (at latitude 66.5°S) or for phenomena, objects, or organisms originating in that area.

Bends Common term for decompression sickness. Humans are subject to this serious disorder when gasses dissolve in their blood during prolonged periods of breathing compressed gas at depth. When the pressure is relieved by the diver's return to the surface, the dissolved gas, particularly nitrogen, forms tiny bubbles in the bloodstream that can block capillaries and cause tissues "downstream" of these capillaries to die.

Before Viewing

☐ In Garrison, scan Chapter 6 (to get a sense of contrast between polar and tropical oceanic conditions), then read pages 152–158 describing those differences in detail. If you are sitting in a warm climate or by a warm fireplace, read pages 45–47 and 81–82 on the history of polar exploration.

☐ In Ingmanson & Wallace, read Chapter 17, "The Polar Seas." Note especially Figures 17.13–17.15 on pages 383–384.

☐ Watch Program 24: "The Polar Seas."

After Viewing

Reread the text sections, being sure you understand the key terms and phrases, and complete the following exercises:

1. Which pole has the more extreme temperatures? (That is, which pole has the higher high temperatures in the local summer, and the lower low temperatures in the local winter?) Why is this so?

2. What is the most critical limiting factor for polar oceanic plants?

3. If a polar day is six months long, why is the peak of polar biological activity only one or two months long?

4. Seals and whales don't get the bends even though they dive much deeper than any humans can with equipment. Why is this so?

5. If June and July are the peak productivity months for the Arctic, are they also the peak productivity months for the Antarctic?

6. The *Nautilus* went to the North Pole in 1958. Why has no one gone to the South Pole by submarine?

7. Can you tell an Arctic iceberg from an Antarctic iceberg just by looking at it? Could you tell the difference if you were given just a core sample to analyze?

8. Do icebergs ever end up in the tropics?

Answer Key for Exercises

1. Perhaps you recall from an earlier lesson that water tends to moderate temperature. The oceans of the world act as a giant "heat sink" to thermostatically control temperature extremes. Because the northern polar region is dominated by the Arctic Ocean, and because the southern region is continental, the thermostatic effects of liquid water are available only in the north. Therefore, the Antarctic is by far the coldest continent. The world's lowest recorded temperature was recorded by Soviet scientists in Antarctica on July 21, 1983. It was an astonishing $-89.2°C$ ($-128.6°F$). On the Antarctic peninsula, the highest temperature recorded was $59°F$ ($15°C$). The average high temperature for the interior of the continent is about $32°F$ ($0°C$). An average summer day is $-20°F$ ($-28°C$).

2. There is plenty of inorganic nutrient material available, and at low temperatures an adequate supply of oxygen is available, so the critical limiting factor is sunlight. Without sunlight no photosynthetic activity can occur; therefore, no primary productivity.

3. The sun must be at a significant elevation over the horizon in order to penetrate the surface of the ocean and be available to phytoplankton. Although direct sunlight returns to the area of the poles themselves only every six months, twilight lasts much longer than that. But this twilight, and the low solar angle that follows it, are insufficient for photosynthesis. Only when the sun is near its highest elevation, around the last of June and first of July in the Northern Hemisphere, does the productivity peak begin.

4. The key to understanding this interesting concept lies in realizing that whales and seals do not breathe supplemental air after they leave the surface. Human divers do. The seals and whales saturate their tissues with oxygen before leaving the surface, and then draw on these reserves at depth. No new gasses are inhaled, so no excess gas can dissolve in the blood. When they return to the surface, there is no excess nitrogen (or other gasses) to leave solution in the blood and block capillaries, so no dangerous symptoms develop.

5. No. The seasons are reversed. December and January are the months of peak productivity in the southern polar ocean.

6. It would be impractical to do so. Whereas the north polar region is covered by an ice cap floating on a deep polar sea, the Antarctic region is underlain by a continental mass. It would be prohibitively expensive to transport a submarine overland to the South Pole.

7. Usually you can. If the iceberg is flat, covers quite a large area, has uniformly steep wall-like sides, and you have reason to believe it is floating rather high in the water, you are probably looking at an Antarctic iceberg. If, however, the iceberg is jagged and pinnacled, colored deep blue-green, and

floating low in the ocean, it is probably of Arctic origin. You might also check your location. The core sample guess would be tricky. The best hint would be the relative "fluffiness" of the ice. Remember, large Arctic icebergs are calved from glaciers and consist of dense, compacted masses of ice. Antarctic bergs are usually less dense because they have not been subject to the compressional forces within a glacier. There are exceptions, of course.

8. Yes. They don't cross the equator (remember the situation of the currents), but they can come to surprisingly low latitudes. One gigantic Antarctic iceberg arrived near the tropics in the 1920s carrying a startled and bewildered (and warm!) group of Adelie penguins. Perhaps the evolutionary precursors to today's Galápagos penguins arrived at their equatorial home by a similar means.

Optional Activities

1. Many fascinating historical accounts of polar exploration are available. Some interesting film footage and still photographs of these expeditions are frequently seen in documentary programs. The diaries and journals of the explorers have been widely excerpted, and accounts of the major expeditions are plentiful. You might especially enjoy the tale of Shackleton's boat journey from stranding on the Antarctic continent to remote South Georgia Island, Nansen's story of the entrapment of Fram in the Arctic ice, and particularly the many analyses of Robert Scott's tragic "Terra Nova" Antarctic expedition of 1911–12. (See the Supplemental Reading for this chapter.)

2. When you visit a zoo or an oceanarium, note the special external adaptations exhibited by pinnipeds and penguins for their rigorous environment.

Self-Test

1. Which of these factors is most severely limiting to polar phytoplankton?
   a. silicates (inorganic nutrients)
   b. water temperature
   c. dissolved oxygen
   d. dissolved carbon dioxide
   e. sunlight

2. Which of these illumination phenomena is found *only* in the polar regions?
   a. continuous sunlight through six months' time
   b. aurora
   c. mirages
   d. ice crystal effects in the upper atmosphere
   e. meteors

3. Table-shaped icebergs extending over broad areas and having flat, sheer walls are characteristic of
   a. the northern ocean area
   b. the southern ocean area
   c. the southern equatorial regions (icebergs take this form only after long immersion in warming water)
   d. both ocean areas
   e. There are no icebergs shaped as described.

4. The zooplankton density during optimal conditions in the polar seas, in comparison with the animal support capabilities of land,
   a. is greater on a pounds-per-acre basis
   b. is about the same order of magnitude as land on a pounds-per-acre basis
   c. is considerably less than that of land on a pounds-per-acre basis
   d. cannot be compared
   e. none of the above

5. The first expedition to reach the South Pole was led by
   a. Peary in 1909
   b. Scott in 1912
   c. Shackleton in 1910
   d. Amundsen in 1911
   e. Byrd in 1923

6. Perhaps the most difficult problem experienced by polar mammals is
   a. mate location and identification
   b. retention of body heat
   c. feeding
   d. navigation
   e. maintaining and defending territory

7. The largest polar mammals are
   a. polar bears
   b. pinnipedians
   c. baleen whales
   d. odontocete whales
   e. krill

8. Which of these polar animals does *not* occur naturally north of the equator?
   a. seals
   b. sea lions
   c. baleen whales
   d. penguins
   e. odontocete whales

9. Which of the following is *most nearly* correct?
   a. Polar species are abundant.
   b. Polar species are rare.
   c. Polar species are rare, but the number of individual organisms of a few of these species can often be abundant.
   d. Polar organisms are often found representing great numbers of species *and* individuals.
   e. Very few individual organisms are found in polar marine and terrestrial environments.

10. Which of the following statements regarding the Antarctic Convergence is true?
    a. It is a place where Antarctic water and temperate water meet in the southern ocean.
    b. It is a spawning ground for polar mysticete whales.
    c. It is located at about 80° south latitude.
    d. It is the farthest place north where icebergs can possibly drift.
    e. It is a source of unusually cold water.

Supplemental Reading

Amundsen, R., *The South Pole*. London: John Murray, 1912.

Huntford, R., *The Last Place on Earth*. New York: Atheneum, 1985. (History of the race for the South Pole. A controversial book "debunking" the popular view of Captain Scott as mythical hero and used as the basis for the PBS/BBC television miniseries of the same name. Truth probably lies somewhere between this harsh view and the more conventional portrait drawn by Elspeth Huxley [see below].)

Huntford, R., *Shackleton*. New York: Atheneum, 1986. (An excellent, complete, and largely sympathetic biography of Sir Ernest Shackleton. Very entertaining, especially if you have a warm fireplace handy!)

Huxley, E., *Scott of the Antarctic*. New York: Atheneum, 1978.

King, H. G. R., *The Antarctic*. Blandford Press, 1969.

Worsley, R. A., *Shackleton's Boat Journey*. New York: Norton, 1977. (This is Worsley's account of the Shackleton expedition of 1914–1916. Worsley was captain of *Endurance*, and it was his seamanship that made the boat journey to South Georgia possible. Worsley died in 1943. This book is very highly recommended.)

Lesson Twenty-five   The Tropic Seas

Overview

Girdling the earth is a belt of incredibly blue water, lush green islands, and vari-colored coral reefs known as the tropics. This region has inspired countless artists and writers, and is blessed with such beauty that expressions such as "tropical island paradise" have become part of our everyday language.

The tropics are the result of the unique relationship between Earth and sun and the tilt of the earth's axis. As the seasons change, the direct or vertical rays of the sun move between the Tropic of Cancer and the Tropic of Capricorn. It is this zone between latitudes 23.5°N and 23.5°S that defines the tropics. Here the sun is high in the sky during all seasons of the year and winter never comes. Here the oceans are heated and develop the special properties that characterize the tropical seas.

The tropics are also within the belt of the trade winds that blow from the northeast in the Northern Hemisphere and from the southeast in the Southern Hemisphere. These winds meet just north of the equator and gently rise in an almost windless belt called the doldrums. The trades are persistent and can bring strong winds and crashing waves onto the windward sides of the islands. The torrential rains of the tropics are also on the windward side, as the moisture-laden air hits the land masses, rises, cools, and releases its moisture. The lee sides, away from the trades, may be dry as deserts, and will support a typical desert vegetation. The steamy, lush green isles of the travel brochures are real, but are usually the small islands or just the windward side of the larger islands.

The oceans of the tropics are a brilliant blue, shining with sparkling clarity. The surface waters are warm, heated by the intense rays of the sun. The currents flow parallel to the equator, driven toward the west by the trades. The beautiful water, unfortunately, is low in the nutrients required to support an adequate plankton crop and stable food webs. The islands are too small to contribute nutrients from run-off, and local upwellings from the small coastal currents are insignificant. Whatever organic material there might be falls through the thermocline and cannot be recycled back to the less dense, warm surface layers. Thus, the lovely, brilliant blue water is actually a biologic desert, a region where the struggle for survival is often intense.

The teeming, colorful marine life usually associated with the tropics is concentrated around the reef communities that hug the shore of the islands and the continents. The great architect and builder of the tropical seas is the coral animal. Smaller than an ant, the individual coral polyps work together as a colony in perfect coordination, creating such elegant structures as the brain coral, elkhorn and staghorn coral, finger coral, palmate coral, and a host of others.

Each tiny coral polyp resembles the common sea anemone of the intertidal zone, and, indeed, they are related. Both belong to the phylum Cnidaria (Coelenterata). Other animals in this group include the jellyfish, the Portuguese man-of-war, tiny hydroids, and many other species. The Cnidaria are all radially symmetrical and have a mouth surrounded by tentacles armed with stinging cells. The corals are no exception.

Coral polyps feed mainly on small plankton, suspended organic material, and can take in compounds dissolved in seawater through their body wall. Corals with large polyps feed on small fish captured by the stinging cells on their tentacles. The flowerlike coral polyps are actually carnivorous animals. But they fill several ecologic niches simultaneously. They are primary producers, primary consumers, and detritus feeders, as well as carnivores. This versatility reduces their dependence on any one food source in a changing and often unproductive environment.

Because the tropical waters have low concentrations of dissolved nitrates, ammonia, and phosphates, most of the reef corals obtain additional nourishment provided by masses of zooxanthellae that live in their cells. These yellow-brown dinoflagellates conduct photosynthesis, and divide and grow within the cells of their coral host. In this classic example of a symbiotic relationship, the corals provide the plant cells with a constant, protected environment and an abundance of nutrients – carbon dioxide, nitrates, phosphates, sulfates, and ammonia – all waste products from cellular respiration. The algae recycle these compounds into new organic material and oxygen. The corals do not digest the plant cells directly, but apparently absorb amino acids, glucose, and other organic compounds resulting from the plant photosynthesis. The algae also contribute to the ability of the polyps to rapidly deposit calcium carbonate required for the growth of the coral skeleton.

Corals of various kinds are found in many areas of the sea, from deep to shallow water, as well as on temperate and tropical shores. However, the true reef-building, or *hermatypic*, corals are limited to tropical waters where the temperature never dips below 18°C and preferably stays between 23°C and 25°C. For optimum development, reef corals also require a good circulation of clean water free of sediment, a firm bottom, moderately high salinities, and plenty of sunlight for the photosynthetic zooxanthellae. Consequently, active coral growth does not occur in water much below 46 m (150 ft), nor are reefs found near river mouths where the water is too fresh and carries suspended sediment.

The reef itself is actually a complex community in which the corals play a major but not exclusive role. The coral skeleton is composed of calcium carbonate that the polyps absorb from seawater and form into crystals for skeletal material. Large quantities of calcium carbonate are added to the reef structure by several kinds of red and green calcareous algae. In some reefs, as much as half the carbonate may have been contributed through the activities of plants. Shells of foraminifera, snails, and clams, and tests and spines of sea urchins and other animals with hard parts of calcium carbonate also become part of the growing reef.

Coral colonies, plus the fishes and invertebrates that live *near, on,* or *in* nooks and crannies of the porous reef structure function as a complete and highly efficient system. Each has its own photosynthetic, herbivorous, and carnivorous components. Critical nutrients are rapidly recycled between the producers and the consumers. Few are lost, and localized productivity is higher than in most ecosystems.

The ultimate shape the reef will take depends on many factors: wave action, currents, nutrients, contributing organisms and, of course, the base or substrate upon which the reef is built. No two reefs will be exactly alike, and there are thousands of reefs dotting the tropical waters. It was Charles Darwin, while serving as a naturalist aboard the HMS *Beagle* in the 1830s, who classified the reefs into three general types and proposed a theory of origin for the ring-shaped coral atolls. The first stage in reef formation is the fringing reef that borders the shores of the volcanic islands. Many Hawaiian reefs are fringing reefs. Barrier reefs are offshore and are separated from the shore by a lagoon. The Great Barrier Reef of Australia is by far the largest single biological feature on earth. It differs from true oceanic island reefs in that it is built upon a continental shelf and will have a different geologic history than an island reef. Atolls are ring-shaped reefs surrounding a lagoon with a few coral islands projecting above the water level.

According to Darwin's theory, the fringing reef forms first around a new volcanic island, which is populated by organisms reaching the newly emerged land as planktonic larvae. As the reef grows, the island sinks, both from the weight of the reef and the island and from movements of the crustal plates. The polyps continue to build the skeletal structure upward at about the same rate as subsidence is occurring, forming the second-stage offshore barrier reef. Eventually the island subsides beneath the sea, but the living corals continue to build their structure upward, and the ring-shaped atoll is left. If subsidence is faster than the rate of coral growth, the structure may descend into the cold dark depths, and the reef community will perish.

The reef community, as efficient as it is, makes up a small part of the tropical seas. Most of the life in these waters is dominated by fierce competition for food, territory, and opportunities to breed and continue the species. In the warm waters, biological and chemical reactions are rapid. Rates of growth, metabolism, and reproduction are all increased. The animals grow rapidly, but do not reach a very large adult size. They will reproduce while very young but will not have a long life span. There are many different species of animals but relatively few individuals of each. Huge schools of fish, for instance, are not characteristic of the tropics. Many tropical animals are brightly colored, to attract a mate, lure a prey, confuse a predator, or for species identification. Venoms and toxins are also common among tropical forms, again for protection and for food gathering.

The fascinating ecosystem of the coral reef is threatened today on certain populated islands. Construction, dredging, and landfills loosen the soil that washes over the fringing reef, burying the delicate polyps. Untreated sewage dumped in bays has killed polyps and other reef organisms, leaving dead reefs populated by only a few hardy sea cucumbers and masses of strange bubble or filamentous algae. The practice of collecting certain animals for their attractive shells has also upset the delicate balance. It has been suggested that some predatory animals, such as the crown-of-thorns sea star, have proliferated when their predators were removed (in this case the large triton snail). This large, bristling, many-armed starfish feeds on coral polyps and will move through a reef leaving a swath of bare skeletons.

For the visitor with mask and snorkel, the coral reef is an oasis, a wonderland in the vast blue desert. The reef is a treasure to be studied and enjoyed, but also guarded as one of the most precious jewels in the sea.

Learning Objectives

After completing the reading assignment and viewing the program, the student will be able to:

☐ Describe the physical characteristics of the tropic seas and relate these properties to biological adaptations for survival.

☐ Explain the relationship between hermatypic coral polyps and their symbiotic algae in terms of mutual benefits.

☐ Describe the coral animal and how it initiates reef construction.

☐ Discuss the reef ecosystem and how it functions.

☐ Compare and contrast the three kinds of reefs and describe the origins of each.

☐ Appraise the threat of destruction of living reefs in terms of natural cycles and human – activities.

Key Terms and Phrases

Tropics The zone lying between the Tropic of Cancer and the Tropic of Capricorn, heated by the vertical rays of the sun. The position or latitude of the tropics is determined by the tilt of the axis, 23.5° from a line perpendicular to the plane of the earth's orbit, and by the fact that the axis does not swing but is in a fixed direction in space. The tropics is the zone of maximum heating of the oceans, of deepest penetration of sunlight, and, in places, of lowered surface salinity due to excess of precipitation.

Cnidaria (Coelenterata) A group or phylum of animals that are radially symmetrical, tubular, or cup-shaped, have tentacles armed with stinging cells (nematocysts), and exhibit one of two basic body plans, the polyp and the medusa.

Polyp One of two body forms taken by cnidaria. The polyp is the attached form in which the base of the animal is fastened to a substrate or sits in a cup of its own making. The free end contains the mouth surrounded by tentacles. The polyp form is that of the coral and sea anemone. The medusa form in which the cup is inverted and the tentacles hang down is free-swimming and is characteristic of the jellyfish, Portuguese man-of-war, sea nettle, and many larval cnidaria.

Hermatypic or reef-building corals Polyps that contain masses of symbiotic unicellular plants called zooxanthellae, which live in their surface tissues. The zooxanthellae need sunlight and clear water to photosynthesize; therefore, the living part of the reef is restricted to the shallow photic zone. Hermatypic corals are found only in tropical regions where the temperature of the water is never below 18°C. Reef corals are more abundant and diverse in the Indian and Pacific Oceans (about 700 species) than in the Atlantic (about 35 species).

Ahermatypic corals Colonial corals that do not form great reefs and do not contain symbiotic algae in their body cells. They can be found in the deep sea as well as in colder latitudes. Ahermatypic corals can be found in reef communities, but do not contribute greatly to the reef structure.

Before Viewing

☐ Review the differences between the polar and tropical ocean areas on page 433 of Garrison, then continue by reading pages 151–156 on "The Tropical Ocean." You may also wish to review the information on Cnidaria on pages 379–381.

☐ In Ingmanson & Wallace, reread "Seamounts, Coral Reefs, and Island Chains," pages 58–60. Reread "Cnidaria" on pages 282–284. Note also the information about coral's internal dinoflagellates on pages 272–275. In Chapter 16, read "Coral Reefs" on pages 354–356. Review the cause of seasons in Chapter 11.

☐ In the Glossary of the text, look up the following words or expressions:
atoll   fringing reef
barrier reef   trade winds
cnidarian   zooxanthella
coral reef

☐ Watch Program 25: "The Tropic Seas."

After Viewing

Reread the text assignment and the Overview section, and complete the following exercises:

1. Describe the series of events that lead to the formation of an atoll, from the birth of a volcanic island to the fully developed ringshaped reef.

2. Describe the physical conditions of the tropical seas, and explain the conditions that contribute to the low biologic productivity. Why is blue water considered a biologic desert?

3. What factors limit the geographic distribution of coral reefs? How do the coral animals

and other animals in the community survive in areas where nutrients are in short supply?

4. What events are taking place today that threaten the survival of coral reefs around the world? Any ideas on what can be done? What will happen to the coral reefs if we experience another Ice Age? The seas will chill below the optimum temperature, and sea level will be lowered 100–125 m (300–400 ft). In fact, what happened the last time? Something else to ponder: Hawaii is moving northwestward with the motion of the Pacific plate. The coral reefs around Hawaii are already near the northern limits of reef survival. What are the long-term prospects for the Hawaiian reefs? (Not good. Don't postpone that visit too long.)

Optional Activities

1. Put on your mask, fins, and snorkel (mentally, of course) and lazily float over a coral reef. Try the reef on pages 426–427 of Garrison or pages 354–356 of Ingmanson & Wallace.

2. Watch out for number 2 in the diagram of the reef in Activity 1. What is this animal and why is it dangerous? To what other animals in the community is it related? (See pages 379–381 in Garrison, or page 301 in Ingmanson & Wallace.)

3. The fishes are brilliantly patterned and colored. For what purposes? Why is this a valuable adaptation, especially for tropical seas?

4. The giant clam, *Tridacna*, number 20 in the reef diagram, has a nasty reputation that is almost totally unjustified. Do you really think it catches and eats unwary divers?

5. Take a close look at numbers 14 and 24. What do these animals do for a living? Take a look at the clownfish and the sea anemones. This is an unusual relationship as anemones usually eat fish this size, but actually protect the little clownfish. What does the anemone get out of this relationship? Unusual relationships or ways of making a living abound in the reef ecosystem. Why is it prevalent in the tropics?

6. Examine the poorly defined colony of animals labeled number 19. These animals have affinities to which other animals? YOU! See Figure 15.20b, page 390 in Garrison, or Figure 14.50, page 303 of Ingmanson & Wallace. What do they have (at least in their larval stages) that you have too? These are tunicates in the phylum Chordata, subphylum Urochordata. The larvae show all of the characteristics of chordates. Eggs hatch into "tadpoles," which have a long tail, gill slits, a notochord, and a nerve tube.

7. Study the reef pictures carefully. Compare the shallow-water communities of the continental tide pools on pages 420–421 of Garrison, or pages 364–365 of Ingmanson & Wallace. What conspicuous members of the intertidal zone are missing from the tropical reef? Check the key to the coral reef habitat and determine if any plants are included in the list.

Self-Test

1. The zone called the tropics
   a. is determined by the gravitational pull of the moon and sun
   b. results from surface circulation and the Coriolis effect
   c. is determined by the tilt of the axis as the earth revolves around the sun
   d. is determined by the high latitude storm regimes
   e. results from monsoon climates

2. The windward sides of tropical islands
   a. generally face into the trade winds
   b. have a very wet climate
   c. have heavy surf and active wave erosion
   d. usually face eastward
   e. all of these

3. Maximum heating of the world oceans occurs in the tropics because
   a. the water is clear and solar rays penetrate deeply
   b. there is little seasonal change; the climate is warm all year round
   c. the sun is always high in the sky
   d. the days are long even in winter
   e. All of these statements apply.

4. The warmed waters of the tropics are noted for
   a. abundant dissolved oxygen
   b. lack of dissolved nutrients
   c. strong open-ocean upwellings
   d. abundant organic material in the surface waters
   e. all of these

5. Of the following statements, which does *not* apply to coral animals? They
   a. are Cnidaria
   b. are radially symmetrical
   c. are carnivores with stinging cells
   d. have a medusa body form
   e. build skeletal structures of calcium carbonate

6. The coral reef community
   a. is made up exclusively of various species of coral polyps
   b. is limited to carnivorous animals
   c. lies within the kelp forest habitat of the tropics
   d. is made up of filter and suspension feeders living off the abundant plankton
   e. is made up of various plants and animals, including photosynthetic producers, herbivores, and carnivores

7. The zooxanthellae that live in the tissues of the coral polyps
   a. provide carbon dioxide and phosphates for the polyps
   b. feed on the tissues of the polyps and are dangerous parasites
   c. are disease-causing and are threatening the reefs of the world
   d. provide additional nourishment and oxygen to the polyps
   e. all of these

8. Coral reefs are found only in areas
   a. of deep water below 200 m
   b. of cold water or in cold currents
   c. where the water has normal or slightly elevated salinity
   d. near rivers or stream deltas
   e. of low sunlight because they need shade to grow

9. According to Darwin, the last stage in the cycle of reef formation is
   a. the fringing reef
   b. the atoll
   c. the algal rim
   d. the barrier reef
   e. the kelp forest

10. Life in tropical seas is characterized by
    a. extreme competition for food, territory, and reproduction opportunities
    b. very few species but large numbers of each species
    c. large adults in each species that reproduce late in life
    d. organisms with extremely long life spans
    e. all of these

11. Many coral reef communities are threatened today by
    a. water pollution off populated areas
    b. dumping of untreated wastes into bays
    c. loosened soil washing in and burying the reef
    d. upsets to the ecological balance by removing certain reef inhabitants for their prized shells
    e. all of these

Supplemental Reading

Goreau, T. F., N. I. Goreau, and T. J. Goreau, "Corals and Coral Reefs." *Scientific American*, vol. 241, no. 2 (1979), 124–136.

"The Great Barrier Reef: Science and Management." *Oceanus*, vol. 29, no. 2 (1988). (The entire issue is about the Reef and organisms that have constructed this remarkable biologic feature.)

Moore, R. E., P. Helfrich, and G. M. Patterson, "The Deadly Seaweed of Hana." *Oceanus*, vol. 25, no. 2 (1982), 54–63. (A mystery story about a deadly cnidarian.)

Ryan, Paul R., "The Underwater Bush of Australia, The Great Barrier Reef." *Oceanus*, vol. 28, no. 3 (1985), 30–41.

Lesson Twenty-six   Mineral Resources

Overview

In the "old days" of oceanography, a few short decades ago, the sea was viewed not only as a last frontier for the adventuresome, but as a vast, almost limitless source of mineral wealth. It was comforting to think that as we learned more about the oceans, we would surely discover the treasure trove of the deep. Actually, of all the mineral resources anticipated from the sea, only two to this date approach and may exceed expectations. They are oil and gas.

Mineral resources on the land have many economic advantages over mineral resources from the sea. Not only are land-based facilities easier to operate, and at a fraction of the cost of offshore facilities, the variety of minerals that can be extracted far exceeds anything we have seen in the sea. On land valuable elements have been concentrated into ores by such natural processes as igneous activity, sedimentation, weathering, even biologic activity. In the oceans the elements are present but in such diluted concentrations that it is not possible to extract them commercially.

A few substances are being produced, however. Salt has been obtained for thousands of years by evaporation of seawater. Today, over 6 million tons of salt per year are obtained from shallow evaporating pans. Ponds at the south end of San Francisco Bay produce about 5 percent of the salt consumed in the United States. Bromine, another element obtained from seawater, is very useful in certain medicines, chemical processes, and in the antiknock formula used in gasoline. About 80 percent of the bromine used in the United States is recovered from ocean water.

The total production of magnesium in this country is also from seawater. Magnesium is similar to aluminum but stronger, and is used

where very strong lightweight materials are needed. Magnesium salts that are soluble and used in medicines are also recovered from seawater. What child has not tasted the miracle worker Milk of Magnesia?

Fresh water from seawater is becoming an increasingly important resource as industrial nations pollute their surface waters, deplete their groundwaters, and need more water for their expanding populations than natural sources can supply. In addition to salt, seawater contains many dissolved substances that are very difficult to separate from the water. Various methods of desalination are in use today including distillation, freezing, reverse osmosis, electrodialysis, and others. All are costly, especially in terms of energy. Solar energy offers great promise, and solar stills are just beginning to be used to recover water for agriculture.

Sand and gravel are not very glamorous marine resources, but they are second in dollar value only to oil and gas. The deposits are widespread, easily accessible, and used extensively in the building industry. Intensive mining, however, may cause environmental problems including beach erosion, changes in benthic communities, and the addition of suspended sediment to the water.

About 90 percent of the total value of minerals from the marine environment is based in petroleum revenues from offshore wells. Crude oil, like water, is one of the very few naturally occurring liquids on this planet. But unlike water, which is a relatively simple combination of hydrogen and oxygen, crude oil is an incredibly complex chemical soup containing tens of thousands of compounds, mostly hydrocarbons.

The formation of crude oil is a complex process that is not completely understood. When petroleum is found, it is almost always associated with marine sediments, suggesting that the original substances were from the sea. Organic matter such as planktonic plants and animals or soft-bodied marine creatures were the raw materials, as land plants were the raw materials in the formation of coal. The organic material apparently accumulated in quiet stagnant basins where the supply of oxygen was low and there were few bottom scavengers. The action of anaerobic bacteria at this stage converted the original substances into simpler organic compounds. Burial of the organic-rich layers, possibly by turbidity currents, may have been the next step. Further conversion of the hydrocarbons must have taken place at considerable depth, on the order of 2 km (1.2 mi) or more, to supply the necessary high temperatures and pressures. Slow cooking under the thick sedimentary blanket for perhaps 2 or 3 million years allowed the chemical changes to occur that produced the crude oil.

At this point in its formation, the crude oil must be able to migrate from its source rock through the overlying formations into a reservoir layer that has abundant pore spaces in which the hydrocarbons can accumulate. The reservoir must also contain geologic traps and an overlying impermeable layer to prevent further migration of the oil. Finding productive structures where this occurred, which are both underwater and deep within the rocks of the continental shelf, is difficult and requires sophisticated technology. Because of the increasing thirst for petroleum products, however, technological development has kept pace.

Drilling for offshore oil is far more costly than drilling on land. The environmental impact of drilling on the outer continental shelf is a problem of concern to both environmentalists and the oil companies. Fortunately, these problems are being studied before extensive drilling in new areas takes place.

Another fascinating resource from the sea with incredible potential, yet undeveloped, involves the manganese nodules of the deep sea–floor. These strange, black objects that litter miles of the Pacific basin have an estimated value of about one thousand billion dollars. Their content of iron, manganese, copper, nickel, and cobalt makes them particularly attractive to industrial nations lacking these crucial materials.

The nodules vary from pea size to over 15 cm (5 or 6 in.), from shapes that are spherical or lumpy to flattened, resembling a hamburger or a summer squash, Some, but not all, seem to have formed around a nucleus such as a shark's tooth or a bit of rock.

The nodules are a deep-sea deposit, abundant on the floor of the Pacific and found in other areas where the rate of deposition of sediments is also very slow. The nodules increase in size at the rate of about 1 mm to 200 mm per *million* years, one of the slowest chemical reactions in nature. From their size we know the nodules have been forming on the seafloor for at least a few million years. During this long period, several meters of muds and clays could have been deposited over them, yet they are not buried. Instead, they litter the surface of the seabed.

Recent studies have shown that the chemical composition and growth rates of nodules seem to be related to the biological productivity of their overlying waters. Beneath the biologically unproductive waters of the mid-latitude open ocean, the nodules are rich in iron and cobalt but grow very slowly (less than 5 mm per million years). In the more productive areas of the sea, such as just north and south of the equatorial Pacific, the nodules are enriched in copper, nickel, and manganese and grow at 10 to 200 mm per million years. It is also possible that the nodules are somehow related to the metal-bearing hot waters emitted at the hydrothermal vents near the midocean spreading centers. At present, we have more questions than answers about the manganese nodules.

The discovery of metal-rich sulfides around seafloor hot springs has spurred tremendous interest among oceanographers during the last few years. These hydrothermal vents have been detected and sampled in the fast-spreading rift zones of the East Pacific Rise near the mouth of the Gulf of California, at the Galápagos Rift, along the Mid-Atlantic Ridge, and on the Juan de Fuca Ridge about 500 km (310 mi) off the coast of Oregon. The minerals are forming in areas of seafloor volcanic activity where heated seawater, carrying large quantities of dissolved metals and sulfur, pours out through vents and fractures. The metals – mainly zinc, iron, copper, lead, silver, and cadmium – combine with the sulfur and precipitate from the surrounding cooler water as mounds, coatings, and chimneys. While these deposits are certainly commercial-grade ores, they are neither large nor extensive. Also, they are subject to solution and oxidation on the seafloor and are not likely to be preserved for long periods of time. Nevertheless, many scientists and institutions are planning a long-range study of the area, including looking for potential ore bodies *within* the seafloor where the minerals are more protected from chemical alteration.

The Red Sea deposits of heavy metals deserve special mention. The Red Sea is another area where crustal plates are diverging and where molten material is close to the surface. Seawater seeps in through faults and fractures, comes in contact with fresh hot basalts, and dissolves certain metals and salts from the hot rock. The recycled water emerges at temperatures of about 100°C and is extremely saline, about 250–300‰. (Remember: average seawater is about 35‰.) Although the solutions are hot, they are denser than normal seawater because of their high salinity and stay on the floor of the Red Sea in deep fault-bounded basins. The metals precipitate within the basins producing metal sulfides, silicates, and oxides of commercial concentration. Recovery of these ores is expected to start soon.

Of great importance to the exploration for mineral resources from the sea is the United States Exclusive Economic Zone, "EEZ." By proclamation of President Reagan on March 10, 1983, the United States claimed sovereign rights and jurisdiction of all marine resources within a region that extends seaward from the coast 200 nautical miles. (Previously designated territorial waters extended only 3 mi from the shore.) The "EEZ" brings within national domain over 3 million square nautical miles of submarine lands, an area 30 percent larger than the land area of the United States and a region of diverse geological and oceanographic settings. On the east coast, for example, the Zone contains a passive continental margin marked by a relatively flat continental shelf, thick sedimentary deposits, and generally lacking active faults. The Zone on the west coast encompasses a narrow active continental margin marked by compressional forces, parts of an actively spreading oceanic ridge crest, major earthquakes, and volcanic eruptions. The Gulf coast presents still another setting, with great salt domes, vast deposits of oil and gas, and urban problems related to settling of land near the Mississippi delta.

The first step in exploring this new region – much of which has not been studied, its resources virtually unknown – has been to map the surface of the seafloor in order to understand the dynamic processes that are shaping our continental margins. The first bottom surveys were conducted off the west coast of the United States because of the diverse energy and mineral resources known to exist there. Over 100 previously unknown volcanoes have been mapped within the EEZ off our west coast. Huge faults, submarine landslides, seamounts, and details of the spreading crest of the Juan de Fuca Ridge are but a few of the features discovered.

Massive deposits of various metallic sulfides are being explored at the hydrothermal vents on the Gorda and Juan de Fuca Ridges. Knowledge of the modern tectonic setting for ore formation has been valuable in locating new ore deposits on land.

Manganese nodules and crusts also occur within the EEZ of the east and west coasts of continental United States, Hawaii, and the Pacific Island territories. Cobalt, present in the manganese crusts, is being studied to determine how the deposit is formed and how it can be retrieved economically.

Many other facets to the EEZ project include research into meteorology, accurate weather forecasting, environmental studies, effects of plate motion on the seafloor, effects of mining on seafloor organisms, and geohazards such as submarine landslides and earthquakes. It is hoped that this new era of marine exploration and research will lead to informed decisions about offshore activities that will benefit not only the United States but all maritime nations.

Maybe the treasure trove is really there, waiting for these new developments in marine technology and international cooperation to tap the riches of the sea.

Learning Objectives

After completing the reading assignment and viewing the program, the student will be able to:

☐ Recognize the difficulties of recovering mineral resources from the sea.

☐ Describe the four resources presently recovered from seawater.

☐ Discuss the origin of petroleum.

☐ Indicate the resources recovered from the continental shelf, the deep seafloor, and within the sea bed.

☐ Describe the appearance, origin, and distribution of manganese nodules; discuss some of the unanswered questions oceanographers are posing about their origin.

☐ Explain common methods of desalination and understand the problems in recovering fresh water.

Key Terms and Phrases

Mineral resources recovered from seawater *Salt*: By evaporation of seawater. Salt is third in value of all minerals recovered. *Bromine*: Useful in medicines and chemical processes. *Magnesium and magnesium salts*: Important metal. *Fresh water*: There are about 1,000 desalination plants around the world today, but the cost is still too great for widespread use.

Mineral resources recovered from the seabed on the continental shelf *Sand and gravel*: Second in value of all minerals recovered. *Gold*: Found off rivers in the continental United States and Alaska, in shallow-water, placer deposits. *Diamonds*: Off the coast of South Africa. *Tin*: From shallow water sands off Malaysia. *Barite*: Used in oil well drilling mud. *Calcium carbonate*: Used for cement. *Phosphate rock*: An important potential resource for use in fertilizers. Occurs on shallow banks off the coast of Florida and in the Southern California Borderland.

Mineral resources on the deep seafloor *Manganese nodules*: The enormous difficulty of mining the nodules economically has prevented their widespread exploitation. In addition, other problems exist, including ownership of the deep seafloor, the effect recovery of the nodules would have on organisms of the deep, and the use of the nodules once they are on the surface. *Metal sulfides deposited around hydrothermal vents*: New vents and hot springs being discovered and undergoing exploration and sampling. *Red Sea brines and metals*: Very rich deposits to be extracted at some future date.

Mineral resources within the seabed *Oil and gas*: Far exceeds the value of all the other minerals put together. Probably one-third of all the oil produced will be from offshore wells. *Coal and iron*: Mined from shafts that reach from the shore to undersea deposits. The largest mining operations are in Great Britain and Japan. *Sulfur and potassium salts*: Recovered through boreholes.

Before Viewing

☐ In Garrison, read Chapter 17, pages 438–446, and pages 456–457. In Ingmanson & Wallace, read Chapter 18, pages 389–402. Note especially the figures relating to petroleum extraction. Read about the Exclusive Economic Zone on pages 460–461 in Garrison, or page 399 of Ingmanson & Wallace. Reread "Manganese Nodules" on pages 131–133 of Garrison, or page 65–66 of Ingmanson & Wallace.

☐ Watch Program 26: "Mineral Resources."

After Viewing

Reread the text section and the Overview being sure you understand the key terms and phrases, and complete the following exercises:

1. Review the minerals recovered from the seafloor and from within the seabed. Why is recovery of minerals from the sea proceeding so slowly? Discuss the problems.

2. What valuable resources are being recovered today from seawater? Why is fresh water becoming increasingly valuable? What are some methods of desalination? What methods hold the most promise for nations in need of water but lacking energy sources?

3. What are the steps leading to the formation of crude oil? What are the requirements for a commercial deposit? (If you are using the Garrison text, see pages 440–442; Ingmanson & Wallace text, see pages 413–415.) What constitutes an ideal offshore location? How is an offshore well drilled?

4. You want to retrieve a good sample of manganese nodules. Where will you look? (See Figure 5.10 on page 130 of Garrison, or Figure 18.6 on page 394 of Ingmanson & Wallace.) You dredge and bring up a full bucket. Describe them. Cut one open. How does it look on the inside? What are some of the questions about manganese nodules that are confounding oceanographers?

Optional Activity

If you live in a coastal zone with offshore drilling platforms, you may visit a rig if you contact the oil companies in charge. They will occasionally conduct a free tour of their facilities as a public service. Sometimes the Oceanic Society or a similar group will arrange a tour.

Self-Test

1. Which of the following statements apply to minerals from the sea?
   a. They are easier to recover than from the land.
   b. They are usually found in concentrated ores.
   c. They are in greater variety than minerals on land.
   d. They are less expensive to recover than from land.
   e. They are more dilute, harder to reach, and more difficult to recover.

2. All of the following resources are recovered from seawater except
   a. bromine
   b. fresh water
   c. salt
   d. cobalt and nickel
   e. magnesium salts and magnesium metal

3. Second to petroleum, the most profitable resource recovered from the marine environment is
   a. gold
   b. diamonds
   c. sand and gravel
   d. coal
   e. phosphate rock

4. Minerals from the marine environment are being recovered in commercial quantities
   a. from seawater
   b. from the seabed on the continental shelf
   c. from rock layers within the continental shelf
   d. from all of the above
   e. from none of the above

5. The formation of crude oil will occur in an environment that provides
   a. quiet basins for accumulation of organic material
   b. little oxygen and the presence of anaerobic bacteria
   c. few bottom scavengers
   d. deep burial by sediment
   e. all of the above

6. Oil and gas in commercial quantities have been found
   a. accumulated against faults
   b. in the crest of anticlines
   c. against the sides of salt domes
   d. in faulted layers above salt plugs
   e. all of the above

7. Manganese nodules are not considered a valuable source of
   a. manganese
   b. barite
   c. cobalt
   d. nickel
   e. copper

8. Manganese nodules can be found
   a. in shallow water on continental shelves
   b. on the land
   c. in small pockets on the seafloor
   d. as widespread black lumps that litter the deep seabed
   e. all of the above

9. For the future, likely places to look for commercial concentrations of minerals will be
   a. in the Red Sea brines
   b. in the rift valleys of the midocean ridges
   c. in hydrothermal vents on the seafloor
   d. in fissures erupting lava and hot waters
   e. all of these

Supplemental Reading

Bonatti, E., "The Origin of Metal Deposits in the Oceanic Lithosphere." *Scientific American*, vol. 238, no. 2 (1978), 54–61.

Hunt, J. M., "The Origin of Petroleum." *Oceanus*, vol. 24, no. 2 (1981), 53–57.

Mottle, M. J., "Submarine Hydrothermal Ore Deposits." *Oceanus*, vol. 23, no. 2 (1980), 18–27.

Oceanus, vol. 25, no. 3 (1982). (The entire issue is entitled "Deep Sea Mining" and is an important and relevant reference on the subject. It is recommended reading for the excellent photographs and articles on all phases of deep-ocean mining.)

Oceanus, vol. 27, no. 4 (1984–85). (The entire issue, entitled "The Exclusive Economic Zone," contains articles on both living and mineral resources of the EEZ.)

Penny, T. R., and D. Bharathan, "Power from the Sea." *Scientific American*, vol. 256, no. 1 (1987), 86–92.

Talbot, C. J., and M. P. A. Jackson, "Salt Tectonics." *Scientific American*, vol. 257, no. 2 (1987), 70–79.

Lesson Twenty-seven   Biological Resources

Overview

For the small bands of people living by the sea two thousand years ago, the oceans provided an abundant supply of fish and shellfish. Ancient middens of clam, mussel, and oyster shells found in many coastal regions today, attest to human use of the sea as a food resource.

With today's increasing populations, the demand for food is growing at an alarming rate. Food production practices and distribution patterns are unable to satisfy population needs, and starvation and malnutrition are major problems in many nations. Will the sea be able to provide enough food for all or any of our nutritional requirements?

As with all marine resources, the oceans do not give up their treasures easily. In general, fish in the ocean are not densely distributed, nor are the various ocean areas equally productive. The open sea and the deeper waters do not have large fish populations. Adult bottom-living fish average about one per square meter; pelagic fish living in the midwaters average about one per cubic meter. The open ocean of the equatorial and tropic zones is also relatively unproductive.

The largest share of the global marine catch, over 68 percent, comes from shallow waters over the continental shelves in the temperate waters of the northwest Pacific and the northeast Atlantic Oceans. The competition for fishing rights in these regions has been intense. Not too long ago, Britain and Iceland were engaged in a "cod war." British trawlers were attempting to fish in the productive waters around Iceland, having pretty well depleted their local stocks. Iceland objected strongly, to the point that shots were fired to protect the resource.

Of the thousands of species of fish and invertebrates living in the oceans, less than one thousand species are regularly caught and processed. Nine major groups supply about 80 percent of the total world catch. The largest harvest is of clupeid fishes, which include the herring, sardines, and anchovies. At one time, anchovies constituted the largest catch of a single species. For a few years, the Peruvian anchovy boats harvested about 20 percent of the *global* fish catch. The El Niño of 1971–72 dropped the catch from about 11 million metric tons in 1971 to about 4.8 million tons in 1972. By 1980, less than 1 million metric tons were recovered, and the recent catch, further decimated by the 1983–84 El Niño, has been less than 100,000 metric tons. In terms of total tonnage of fish caught, Peru slipped from first place to sixth. Japan leads the world today but with fish caught mostly outside its fishing boundaries. Of course, not all of the catch is for human consumption. A sizable portion is used for pet food, livestock fodder, fishmeal for chickens, oils, and fertilizers. In fact, when the catch of anchovies goes down, the cost of fried chicken goes up.

The clupeid fishes (herring and others) still make up over 23 percent of the global catch. Cod, haddock, hake, and their relatives, and the flatfishes (such as halibut, sole, flounder, and turbot) contribute about 12 percent of the total tonnage.

During the 1960s, increased fishing, primarily by fast long-range vessels, produced an increase in the tonnage of the world catch. In the 1970s, the total catch did not increase in spite of the continued intense fishing. In the 1980s, the tonnage seems to be increasing again, probably due to greater demand and higher prices for fish products. New at-sea processing has resulted in better quality seafood reaching the markets. Also, the health benefits of eating fish are now widely touted, increasing the demand and consequently the price.

It is difficult to evaluate with certainty the significance of these variations in total harvest. Perhaps as some species get fished out, other species will expand to fill the ecologic niches. Also, fish-finding technology is improving. In any case, the catch is not increasing at the same rate as the growing human population. With present fishing practices, the future looks dim for the more desirable species.

An interesting new fishery in the Antarctic Ocean is being considered. The Antarctic is highly productive because dissolved oxygen and organic and inorganic nutrients are abundant in the cold upwelling waters. In the long days of summer there are vast blooms of phytoplankton, which are the basis for the Antarctic food chains. The diatoms are consumed by small euphausiid crustaceans called krill, the "whale" food of Norwegian whalers. The abundance of krill is legendary. Krill harvesting and processing were pioneered by the Soviets and Japanese in the 1960s. Today, the Soviets are producing krill butter and krill cheese spread, as well as poultry and cattle feed. The Japanese are experimenting with fish cakes and a type of fish sausage of krill paste thickened with algae gel. Krill is also sold raw fresh-frozen.

The great danger in utilizing krill lies in the fact that in the Antarctic ecosystem many large predators depend on a single kind of prey. Krill makes up about 80 percent of the diet of blue, fin, humpback, and minke whales and perhaps 100 percent of the diet of the crabeater seal. One effect of intensive whaling has been the proliferation of seabirds, penguins, squid, fish, and benthic organisms, attributed to reduced whale populations and the increase in krill. If krill is harvested at the rate anticipated by certain nations, will the blue, fin, humpback, and right whales – all endangered species, all depending on krill – ever recover to a viable population level? Even the sperm whale is threatened because its prey is squid, which in turn feeds on krill.

This one example illustrates the complexity of harvesting biological resources from the sea. Humans have never shown that they are prudent predators. The motivating force is quick financial return, even if it means depleting a stock and disrupting the equilibrium of a fragile ecosystem.

When it comes to harvesting the sea, humans are still primarily hunters, not too far removed from their Paleolithic days. We still catch our prey wherever we find it and do little to farm, ranch, or herd desirable species. Most fish farming or aquaculture today involves freshwater organisms and, in spite of both need and promise, this effort is advancing rather slowly. One exception is China, where freshwater fish production exceeds all other nations. Their aquaculture program provides the large Chinese populace with an abundant supply of high-quality, extremely popular animal protein at a reasonable cost.

The problems of marine mariculture are many. The coastal waters where the fish ranches are most likely to be located are polluted, especially near urban areas. Therefore, tanks or structures must be used where the environment can be controlled. Disease is an important problem in artificial environments, and some desirable species seem especially vulnerable. Another factor to consider is expense: The high cost of feed leads to the high cost of fish, since most are fed special high-protein diets to increase the rate of growth. Even the process of controlling temperature and salinity becomes expensive because of the labor involved.

The life cycle of some of the organisms makes mariculture risky. Crustaceans, such as lobsters, molt and shed their external skeleton from time to time as they grow. During this period, they are very susceptible to disease, temperature changes, and particularly cannibalism. Any project involving the commercial growth of lobsters must use warmed water, about 22°C, which is warmer than usual seawater. Also, the animals must be kept in separate containers for at least part of their life cycle. Since lobsters and shrimps are high-priced luxury items, the risk seems worth taking.

An interesting experiment in abalone farming is being tested under an offshore oil well drilling platform off southern California. The abalones are doing well in their suspended cages, but are very slow growing and must be supplied with food, making the project very costly.

The most successful mariculture projects so far in the United States have been with oysters and salmon. Oysters, as well as mussels and clams, are filter feeders that take their food out of the water. For this reason, they are less costly to raise. The Pacific oyster industry currently produces over 6.5 million pounds of oyster meat annually, a figure that is sure to increase.

The story of ranched salmon is one of the most interesting in mariculture lore as it is based upon the migratory habits of the fish. Wild salmon will swim upstream, sometimes many miles from the sea, to spawn. They hatch in fresh water where they spend at least their first few weeks. Then they migrate to the sea to live and grow. After about two or three years, the salmon return to their native stream to spawn again. Atlantic salmon return year after year while Pacific salmon return once, only to die after spawning. This homing instinct of migratory salmon is legendary and is based on their ability to recognize by smell the unique chemical signature of their native stream.

Attempts to artificially rear salmon started in Germany in 1763. In the United States the effort to culture salmon began around 1804 for the Atlantic salmon and in 1870 for Pacific salmon. In modern salmon ranches, eggs are hatched in carefully maintained ponds, tanks in a stream, or some other body of fresh water. The young salmon are carefully nurtured before being released into the stream to find their way to the sea. These ranched salmon are homing in large numbers back to release sites in North America, Japan, and Siberia.

Salmon have also been successfully transplanted to Chile, and a new experimental stock is being raised to feed on summertime concentrations of krill at the margins of Antarctica. Those introduced into the Great Lakes have adjusted to that freshwater habitat and are providing a valuable new fishery for the region. Along the rocky coasts of Scotland and Norway, fish farmers are raising salmon and rainbow trout in blocked-off sea lakes, fjords, and floating pens. In addition, over the past 20 years the Japanese have had great success in producing chum salmon at hatcheries. In fact, nearly all the 20 million or more chums harvested in Japanese territorial waters in 1982 are of hatchery origin. Until recently, salmon and wild trout were declining to dangerously low levels due to the decline of streams, overfishing, and loss of habitat. Today, scientists may have reversed the trend through careful ranching and breeding programs. It is expected that the stocks of these fish will once more be abundant.

A small and under-utilized resource, the shark fishery, is very limited in the United States. People are just beginning to recognize the value of this delicious and still rather inexpensive food. It has only been in the past few years that shark has appeared in the supermarkets, particularly mako, hammerhead, and thresher.

Ironically, certain fast-food chains have been using shark for years, very successfully and very quietly. Shark is also used for animal food; cats of the world thrive on it.

Tuna, a favorite and the mainstay of the salad and sandwich crowd, is a relatively small fishery in terms of the world market. Tuna can no longer be caught in coastal waters, and the expense of bringing the fish to market has made it costly.

The commercial harvesting of marine mammals must be included in any discussion of biological resources. Most marine mammals, particularly whales, are now protected by international agreement, and since 1987 all commercial whaling has ceased. Iceland and Japan still harvest whales for "scientific purposes," but the numbers are relatively few. The United States does allow certain native Alaskans to hunt whales, supposedly for subsistence.

The fur seal industry, however, is still actively pursued by several nations, although the United States terminated its fur seal industry in 1986. Every year, in spite of the tremendous outcry, the killing of the baby Harp seals by Canada and Norway still takes place. They claim that their controlled harvests do not decimate the breeding populations.

It has been estimated that at least 50,000 northern fur seals and thousands of sea birds perish in drift nets used by the Japanese in salmon fishing. The decline in the fur seal population may be related to the use of these nets rather than to the controlled harvesting of the seals.

Porpoises are inadvertently taken in purse seine nets used by commercial tuna fishermen. In the United States, the tuna fishing boats are able to release most of the captured porpoises from the nets, but fishing practices of other countries caused an estimated 100,000 porpoise deaths in 1987.

For the enterprising scientific investigator, there are many exciting possibilities and opportunities in the fields of mariculture, fisheries management, marine mammal management, and processing and marketing of food products from the sea. Let us hope that with proper knowledge, we will learn to grow our own biological resources and take some of the pressure off the wild marine organisms. Let the creatures of the sea return, if they still can, to their natural abundance, and let us share rather than destroy the bounty of the sea.

Learning Objectives

After completing the reading assignment and viewing the program, the student will be able to:

☐ Describe the effects of uncontrolled harvesting of our marine biological resources.

☐ Define the term *maximum sustainable catch* and identify the limited value of the concept.

☐ Compare and contrast the successes and the problems associated with aquaculture and mariculture.

☐ Discuss the benefits and dangers of establishing a large-scale krill fishery in the Antarctic.

☐ Discuss the life cycles of Pacific salmon and relate these cycles to successful ranching techniques.

☐ Understand the concepts of endangered species; evaluate the present conditions affecting survival of major marine organisms studied in this course.

Key Terms and Phrases

Maximum sustainable catch The maximum weight of any species that may be harvested year after year without causing a long-term decline of that population. The concept of maximum sustainable catch is limited in value when applied to a single species, since most have strong interactions with other species. A whole ecosystem must be considered.

Overfishing Harvesting of a species beyond its capability to reestablish its natural population. The numbers will decline each year until it is no longer of commercial value. Overfishing, especially during the El Niño years, is one of the factors that led to the decline of the Peruvian anchovy, which has not yet returned to its former level. It is believed, however, that overfishing has never driven a species of marine fish to extinction.

Endangered species A population of organisms so reduced that it is in danger of extinction. *Marine fish*: None that live exclusively in ocean waters are on the endangered list. *Marine invertebrates*: None are listed as endangered, although many habitats have been destroyed and pollution of coastal and estuarine waters has severely limited certain valuable species. *Marine reptiles*: All living species of crocodiles and many marine turtles are endangered. Turtles are hunted for food in certain countries and crocodiles have been hunted almost to extinction by sportsmen. *Marine birds*: They are severely stressed by pollution, loss of habitat, and by human activities in the coastal zone. At least 31 species are now considered endangered. *Marine mammals*: Certain forms are still on the endangered list although commercial harvesting has been discontinued. The stocks of several whale species are extremely limited. The great blue whale, whose numbers are estimated to be about 1,000, may be functionally extinct.

Aquaculture The growing or farming of plants and animals in a water environment under controlled conditions. Aquaculture is an important industry in China, Japan, the Soviet Union, and other countries. Trout and catfish are successfully grown in the United States, while carp, milkfish, and certain other species have been farmed elsewhere.

Mariculture The farming of *marine* organisms, usually in estuaries, bays, near-shore environments, or specially designed structures using seawater. Several species of fish, including plaice and salmon, have been grown commercially. Invertebrates farmed include shrimp, mussels, oysters, and abalone, as well as several kinds of edible seaweed. In Japan, attempts to ranch yellowtail and tuna met with limited success, but Japan still produces many other marine organisms and leads the world in mariculture.

Krill Small shrimplike crustaceans of the genus *Euphausia*, abundant in the zooplankton of the Antarctic seas. *Euphausia* is the principal food source of many baleen whales, seals, seabirds, penguins, fish, and squid. Krill is now being harvested at a rate of about 400,000 tons per year. The taking of krill has serious implications for the depleted stocks of baleen whales and other creatures of the south polar seas.

Before Viewing

☐ In Garrison, read Chapter 17, pages 446–456, "Biological Resources." Reread the material on trophic relationships on pages 349–351 (and note the food web on page 351). Note also the information on krill on pages 366–367.

☐ In Ingmanson & Wallace, reread Chapter 14, pages 296–297 on krill. In Chapter 15, "Vertebrates," review the sections regarding the commercial importance of fishes. Read about the biological resources on pages 402–414.

☐ Consider the problem of overfishing krill. Note the central position of krill in the food chains represented in the texts. If krill were overfished, what would happen to the food chains?

☐ Refer to Lessons 16, 17, 18, and 19 in this study guide as a review of marine organisms, thinking of them in terms of biologic resources.

☐ Watch Program 27: "Biological Resources."

After Viewing

Reread the text references and the Overview section being sure you understand the key terms and phrases, and complete the following exercises:

1. Define maximum sustainable catch (yield) and explain why this concept has limited value.

2. If you are using the Ingmanson & Wallace text, study Table 18.5 on page 405, which shows the leading fishing countries of the world. Which nations account for the numbers for the Northwest Pacific and for the Southeast Pacific? See Figure 18.19 on page 407. From the size of the catch, can you tell if the Peruvian anchovy returned by 1985? What was the effect of the 1983 El Niño? The combined effects of El Niño and overfishing are dramatically demonstrated.

3. Discuss the problem connected with obtaining adequate protein from marine resources. Include such items as marine productivity efficiency, distribution of fish, effects of natural cycles, and effects of overfishing on an ecosystem.

4. What are krill? What uses can be made of this organism? What are some of the dangers of developing a major krill fishing industry in the Antarctic?

5. Compare and contrast the advantages and problems of aquaculture (freshwater ponds) and mariculture (farming the sea).

Optional Activities

1. In the frozen food case at the market, read the contents of different fish products, especially packages that do not identify the species in the title (fish cakes, fish sticks, fish fillets, fish kabobs, and so on). What fishes are most commonly used? Why don't the packagers identify the fish by name?

2. How would you feel about eating krill sausage? Would you prefer eating krill products directly, or having the little crustaceans made into chicken feed and processed through "The Colonel"? Food preferences often are traditional or cultural and have little to do with nutrition, flavor, or availability of the food resource.

3. Try some unusual seafood to expand your gustatory horizons. Taste shark, which is similar to swordfish and is less expensive, Be sure it is very fresh. Then try squid, which is prized in many countries and is gaining acceptance in the United States, particularly in specialty or ethnic restaurants. Try some at home. If you live near the coast, go to the fishing docks and examine the day's catch. *Try something new.*

4. See "Management of Multispecies Fisheries" in the July 20, 1979, issue of *Science* for a technical but understandable analysis of fisheries, in particular the Antarctic krill-fishing industry.

Self-Test

1. The best fishing grounds are located
   a. in the middle of the open ocean
   b. in temperate waters over continental shelves
   c. in tropical waters of the deep sea
   d. in equatorial deep waters
   e. anywhere because fish are migratory

2. The world catch of fish is made up of what kinds of organisms?
   a. thousands of different species of fish
   b. primarily invertebrates, such as lobsters and oysters
   c. mostly herring, anchovies, and sardines
   d. mostly cod, hake, haddock, and herring
   e. mostly tuna and salmon

3. The tonnage of fish caught has changed in what way?
   a. It has been climbing steadily since the 1950s.
   b. It has declined drastically since the 1950s.
   c. It has stayed at the 1960 level.
   d. It increased steadily to about 1970, then declined.
   e. It has increased somewhat since about 1980 due to better fishing boats and increased demand.

4. Krill is important to the Antarctic ecosystem
   a. because they are photosynthetic and are at the base of all food chains
   b. because they are very rare and must be preserved
   c. because they provide food for many large animals in polar seas
   d. because they have been overfished to the point of extinction
   e. because they can be easily grown in ponds

5. Fish that are brought in for processing
   a. are sold fresh or fresh-frozen
   b. are canned
   c. are converted into livestock fodder or chicken feed
   d. are used for fertilizer
   e. all of the above

6. Intense whaling in the Antarctic has affected the ecosystem in which way(s)?
   a. increased the numbers of krill
   b. increased the numbers of seabirds and penguins
   c. decreased the numbers of blue, fin, humpback, and right whales
   d. increased the numbers of seals
   e. all of the above

7. Which of the following statements relate(s) to mariculture as a process for supplementing protein in our diets?
   a. It is used primarily for raising lobsters and high-priced food products.
   b. It is used primarily for ranching large fish such as tuna, yellowtail, and halibut.
   c. It has been most successful in waters near urban areas where markets are nearby.
   d. It has been successful in culturing oysters and ranching salmon in the United States.
   e. It has been unsuccessful in most countries, particularly the Orient, due to lack of technology.

8. Studies of the efficiency of food production in the sea from one trophic level to the next indicate
   a. that it is a very efficient process, with about 95 percent transfer from one level to the next
   b. that it is a moderately effective transfer of about 50 percent of the biomass
   c. that it is an inefficient method; only about 10 percent of the energy and biomass are transferred
   d. that the process is too complex to be measured
   e. that the rate of fishing affects the energy transfer

9. Which of the following statements is true?
   a. Endangered species of marine organisms include most invertebrates except those raised in mariculture ponds.
   b. Large predatory fish, such as tuna, sharks, and marlin, are the only endangered species of marine organisms.
   c. Seabirds are not endangered since they can fly away from inhospitable habitats.
   d. Certain reptiles, such as crocodiles and marine turtles, are among the endangered marine organisms.
   e. Marine mammals are abundant in all oceans and, therefore, cannot be considered endangered.

10. The great change in the world fishery that occurred in the early 1970s, both in total tonnage and in rank of fishing nations, refers to
    a. the collapse of the anchovy population and drop in total world catch
    b. the year of devastating El Niño off the west coast of South America
    c. the crash of the Peruvian fishing industry
    d. the ascent of Japan to first place as a fishing nation
    e. all of the above

Supplemental Reading

Donaldson, L. R., and T. Joyner, "The Salmonid Fishes as a Natural Livestock." *Scientific American*, vol. 249, no. 1 (1983), 50–58. (An interesting and informative article about these remarkable fishes.)

Faulkner, D. J., "The Search for Drugs from the Sea." *Oceanus*, vol. 22, no. 2 (1979), 44–50.

"Harvesting the Sea." *Oceanus*, vol. 22, no. 1 (1979). (The whole issue is devoted to biological resources. This is a very important and informative journal, well worth reading, with nine articles by well-informed writers.)

Hryniewiecka, K., *Let's Cook Squid the European Way*. University of California Sea Grant Marine Advisory Program, 1976.

Oceanus, vol. 27, no. 1 (1984). (This issue, entitled "Industry and the Oceans," covers topics such as salmon ranching, kelp forests, genetic engineering in the marine sciences, and so on. Very interesting and relevant material.)

Peterson, S., and L. J. Smith, "Pitfalls in Third World Aquaculture Development." *Oceanus*, vol. 25, no. 2 (1982), 31–39.

Lesson Twenty-eight   Marine Pollution

Overview

Contamination of the world oceans affects many areas of present research and exploration of the marine environment. There are few pristine beaches or shorelines left to study, and few sea creatures are completely free of the effects of ocean dumping.

It is not a simple matter to define *pollution*. One way to approach the problem is to consider pollution as any additive placed in the ocean that changes the quality of the water or affects the physical and biological environment of the sea. It is not known to what extent the ocean is already polluted. By the time the first oceanographers began widespread testing, the Industrial Revolution was two centuries old and changes were well on their way. It is sad to contemplate that we will never know what clean seas were like, or what remarkable plants and animals may have vanished as a result of human activity.

The rate at which the ocean is changing, even the direction in which some of the cycles are moving, is difficult to determine except in very local areas. The effects of some of the additives may not be felt for many years. To compound the problem, additives that are extremely toxic, even lethal, to certain organisms may function as nutrients to others. Their impact cannot always be predicted. As a result, oceanographers vary widely in their opinions about what is happening to the seas and what to do about it.

Sources of ocean pollution are numerous, including industrial and domestic wastes, fertilizers, plastic residues, atmospheric gases, pesticides, and oil spills. Disposal of chemicals in the sea is particularly detrimental because of the wide variety of substances that are dumped and the far-reaching impact this has on the environment. The case of the vanishing pelican is a classic example.

California pelicans began disappearing in the early 1960s. After intensive scientific detective work it was found that these birds were producing eggs with thin shells lacking in calcium carbonate. The eggs broke easily, no chicks were hatched, and the nests were eventually abandoned. The trail led investigators to the chlorinated hydrocarbon pesticide, DDT, which was in the water absorbed by plankton. DDT accumulated in fishes that fed on these microscopic organisms and in the pelicans that fed on the fishes. The whole food chain was contaminated, but only the higher members were strongly affected. A chemical interaction between the DDT and the birds' calcium-depositing tissues prevented the proper formation of egg shells. The identification of the culprit brought a happy ending to this case, for DDT is now banned in the United States. The pelicans slowly returned home, and the ungainly but charming chicks are seen once more in the nests.

Further research to date suggests a build-up of chlorinated hydrocarbons and PCBs (polychlorinated biphenyls) may even play a role in the behavior changes and declining fertility levels of some populations of pinnipeds (seals and sea lions) on islands off the California coast. These investigations are continuing.

Most startling is the seemingly anomalous rise in levels of DDT, which have been observed in the last few years in waters off California and in the Gulf coast area. Since DDT was banned in the United States, levels had been dropping. One possible explanation for the new threat, and the most likely, is the importation of illegal DDT for use on American crops. The effectiveness, low cost, and stability of DDT as a pesticide make it attractive to the farming industry, but the short-sightedness of the users is unfortunate. It seems that the already stressed populations of birds

159

and mammals may be required to undergo yet another siege of poisoning.

Few tales of marine pollution have happy endings. For the people of Minimata Bay, Japan, who were poisoned by mercury released into the sea from a nearby factory, it is a continuing horror story. Many people in the village lost their lives, but many are still there, suffering from brain and neurological defects including insanity and other severe and handicapping ailments. The discharge of mercury wastes has been stopped, but the bay, which had been the source of food for the poor, is completely unusable for fishing and clamming. This situation will not change for years to come; mercury is not easily removed from the sediments. Once it has contaminated an area, it has a long residence time.

Of course, mercury pollution is not confined to bays. Methyl mercury and other forms of the metal have contaminated fish off the California coast. Pacific mackerel and tuna were found in the early 1970s to contain levels of mercury sufficiently high to prevent their shipment across state lines for the purpose of sale. The response of fishermen was not to demand a cleanup of the pollution sources, but rather to petition for a new, higher limit for mercury in fish shipped between states for human consumption! Though this sounds threatening, it may, in fact, have been a reasonable move. Most of the mercury in the oceans occurs naturally. It was discovered that tuna caught around the turn of this century had an average of 0.38 parts-per-million (ppm) mercury content compared with 0.31 ppm for fresh tuna caught in 1971. Obviously we need to learn more about the origin and toxic effects of mercury in fish and adjust our federal standards accordingly. As is so often the case in environmental issues, things are almost never as simple as they first appear.

Other heavy metals, such as lead, are turning up in increasing amounts in tissue samples of many forms of marine life. The prime source has been tetraethyl lead, which is used in automotive and aviation gasoline. Many other chemicals are released with industrial and domestic sewage. In fact, much of the waste waters entering the sea have had little prior treatment. Trading a polluted river or ocean for a sanitary city, by the way, is an ancient method for keeping urban areas clean – anyone hosing down a driveway knows all about this. The old phrase "the solution to pollution is dilution"

was probably valid when the human populations of the earth were small and chemical processes were primitive. But today the quantity and quality of material spewed out by our coastal complexes is far more than the oceans can safely assimilate.

The actual impact of sewage outfalls is difficult to assess because sewage itself is a complex and variable mixture. The effluent may wander away from its outlet, circulating with the currents to beaches halfway around the world. Or sludge and other insoluble residues may stay near the outfall and carpet large areas of the seafloor, usually with detrimental effects on the benthic communities. Forests of giant kelp bordered the California coast at one time, providing not only a habitat and food for many fishes and invertebrates, but also a valuable resource for a variety of products. The forests have all but disappeared off the coasts where large sewage outfalls are located.

The search for causes of this marine disaster has stirred up something of a scientific controversy. Some authorities suggest that the decline is part of a natural cycle. Kelp is a cold-water plant, and there has been a slight warming trend in the oceans in the last decade. Other theorists implicate the sludge from the outfalls, which coats the rocks and prevents young kelp plants from attaching their rootlike holdfasts. Heavy metals and chemicals in the effluent may also be harmful to the young kelp plants. Though evidence suggests that this sewage sludge can be deleterious to resident populations, the high cost of storing, drying, and disposing of this material on land by burial is currently prohibitive.

The most likely explanation involves the proliferation of spiny sea urchins noted near the outfalls. The young urchins apparently utilize the nutrients in the effluent. Many more grow into adults than the habitat can support. The hungry urchins that normally feed on kelp and organic debris attack the living plants and decimate the forests. It has been suggested that sea otters be brought down from northern California to keep the urchin population in check. The idea was strongly opposed by abalone fishermen because the otters have a greater fondness for abalone than for urchins.

Urchins are only one of the benthic organisms apparently subsisting on nutrients provided by the sewage. Large communities of fishes and invertebrates living around the outfalls, many of which are not those usually found in the area, are adapting to this new food supply in their environment. There is an indication, however, that certain of these fishes show a higher incidence of fin rot and tumors than normally found.

The most widely publicized form of pollution is the release of petroleum into the marine environment. Oil seeps are a natural phenomenon. Some have been leaking large quantities of oil into the sea for centuries. The sources of oil in the oceans, however, have increased greatly in recent years due to extensive shipping of petroleum products, offshore drilling on the continental shelves, near-shore refining, and street runoff carrying waste oil from automobiles.

Oil, especially after refining, is toxic to marine organisms. It forms a layer on the surface of the sea that prevents diffusion of gases and decreases the sunlight available to the waters below for photosynthesis. Oil is particularly harmful to seabirds and may have long-lasting effects on seafloor communities.

A new approach to the problem that has been proposed accepts the fact that oil spill accidents are inevitable and spills will continue to happen with increasing frequency. With proper preparedness, action can be taken not only to prevent and contain the spill, but also to protect particularly sensitive areas rich in marine life. The effects can be minimized, but only through the concerted efforts of government agencies, drillers, refiners, and tankers.

The disposal of nuclear wastes at sea raises yet another dilemma. Although it may be cheaper in the short run to dump radioactive objects and waste into the ocean, the results of our short-sightedness could return to haunt us. Quick dumping could be false economy. Disposal of wastes in this way may be safe, but we do not yet have the data base to allow informed and responsible decision making.

Some nuclear material has already been dropped at sea (near the Farallon Islands off the California coast at San Francisco, for example), and other proposals for ocean disposal have been made. One recommendation, since withdrawn by the U.S. Navy, would have sunk radioactive nuclear submarine parts into deep basins off California. No one could predict the results of these actions.

Even more difficult to predict are the worldwide consequences of actions that, until recently, were not considered detrimental. "Harmless" man-made chemicals released into the atmosphere, primarily chlorofluorocarbons (CFCs) used as refrigerants and in insulating foams, have depleted the earth's protective layer of atmospheric ozone by about 2.3 percent over most of the United States since 1969. Significantly larger decreases have been noted over Australia and New Zealand, and declines to an alarming 50 percent have been observed near the South Pole. Stratospheric ozone intercepts high-energy ultraviolet radiation from the sun. Every 1 percent decrease in atmospheric ozone is accompanied by a 5 percent to 6 percent increase in skin cancer development due to increased ultraviolet radiation. The incidence of malignant melanoma, the most dangerous form of skin cancer, has nearly doubled in the last two decades. The increased ultraviolet light can also kill sensitive planktonic marine food webs. Even land plants such as soybeans have been subjected to sunburn that decreases their yields. NASA has estimated that, if CFC production continues at 1980 levels (less than today's production), the amount of ozone depletion worldwide will be 10 percent over the next century. Some researchers say this estimate is much too low. A sense of great urgency surrounds ongoing research to find safe substitutes for CFCs.

Another planetary problem is the increased rate of global warming due to the "greenhouse effect." Light easily passes through the glass windows of an enclosed car or greenhouse, is absorbed by surfaces inside, and is converted to infrared radiation and heat. Glass in a greenhouse is not transparent to infrared radiation so the energy is trapped and the internal temperature rises. In the case of Earth, carbon dioxide and water vapor take the place of glass – heat that otherwise would be released from the planet is retained by re-radiation from these gases. The human need for quick energy for industrial growth has resulted in a vast amount of carbon dioxide being injected into the atmosphere from the combustion of fossil fuels, especially since the beginning of the Industrial Revolution. Increased carbon dioxide in the atmosphere is thought to have raised the earth's surface temperature by about 10°C (18°F) above base levels. One estimate calls for another 4°C rise in average global temperature within the next 100 years! Such a rise in temperature would result in a rise in sea level of between 2 and 7 m (7 to 23 ft) worldwide after collapse of a portion of the west Antarctic ice sheet, a supposition that calls into question the long-term investment value of shore property! At least one prominent researcher considers the great drought of 1988 an early effect of accelerated greenhouse heating. This drought cost $15 billion, the costliest natural disaster in America's history. The climate changes of the near future will be even more profound.

Another example of the insidious nature of seemingly harmless materials is the fate of plastic debris in the marine environment. Americans generate 133 million tons of plastic waste, about 1,100 lb per person, each year. By the year 2000, plastics in general will account for nearly 10 percent of the entire solid waste mass. Because the ocean is still treated as the ultimate sink, much of this waste plastic finds its way to the sea. A 1987 survey by the Woods Hole Oceanographic Institution found that off the northeast coast of the United States each square mile of ocean surface has more than 46,000 pieces of plastic floating on the surface, including ropes, fishing nets, plastic sheeting and bags, and granules of broken plastic cups. A staggering 100,000 marine mammals and 1 million seabirds die each year after being caught in or ingesting plastic debris! Sea turtles mistake plastic bags for their jellyfish prey and die from intestinal blockage. Seals and sea lions starve from being muzzled by six-pack rings, which also strangle fish and seabirds. Adding ingredients to plastics to aid in their decomposition would add only 5 percent to 7 percent to their cost, a price rise as yet unacceptable to industry. Recycling would be a much better solution than dumping.

Perhaps the most ominous environmental issue of all is the planetary consequence of even a "limited" nuclear war. Studies by a number of scientists of many nations have shown that the injection into the high atmosphere of huge quantities of smoke, ash, and dust from the explosion of nuclear bombs would trigger a "nuclear winter" of unpredictable and unprecedented proportions. The fine particulate material resulting from the explosions would reflect and scatter sunlight, probably change atmospheric circulation patterns, perhaps trigger giant storms, lead to crop failure and famine, and disturb the earth's delicate ecosystems on land and in the sea.

It is useful to realize that we are the *last* generation to live as all other generations of humans before us have lived – to waste resources, throw things away without thought to recycling, live a "dump and go" life, squander energy and fresh water, ignore subtle evidence of negative environmental effects as if they would simply go away. Population pressures are increasing, especially in the desirable coastal zone. By 1990 an estimated 75 percent of all Americans were living within 50 miles of a coastline, including the Great Lakes. Another billion people will be added to the world's population through the next decade, 92 percent of them in the world's poorest countries. The trials ahead will be severe.

We all have an obligation to become informed on pollution issues that affect the earth and its oceans. Often obvious answers are misleading. Much research and work are needed to give reliable insight into the many difficult questions that confront us. The trade-off between financial and ecological considerations is often strongly tilted in the direction of immediate gain. We need to invent new methods of prevention and protection and to strive toward the refreshening and rejuvenation of our sadly contaminated ocean. Education is the first step!

Learning Objectives

After completing the reading assignment and viewing the program, the student will be able to:

☐ Explain the problems that exist in determining the level of pollution in the ocean.

☐ identify the sources of pollution in the marine environment.

☐ Discuss some of the effects of specific pollutants, such as DDT and mercury.

☐ Analyze the kelp problem, taking into consideration the value of kelp both to humans and to the habitat and the various theories that are offered to explain the loss of the kelp forests.

☐ Describe the sources and effects of petroleum in the marine environment.

☐ Describe recent human-induced changes in the atmosphere and discuss the implications of these changes for the ocean and for humanity as a whole.

Key Terms

Pollution In the oceanic context, any additive to the ocean that changes the quality of the water or affects the physical and biological environment of the sea.

Thermal pollution The release of heated water into the oceans from nuclear and conventional power plants. The higher-than-normal water temperatures affect metabolism, growth rates, breeding cycles, and feeding habits of many organisms.

Greenhouse effect Trapping of heat in the atmosphere. Incoming light penetrates the atmosphere, but much of the longer wavelength infrared radiation cannot leave because it is absorbed by gases in the atmosphere and re-radiated toward the surface. Surface temperature rises as more of these gases enter the atmosphere. Carbon dioxide makes the greatest contribution to heating of any "greenhouse gas."

Before Viewing

☐ In Garrison, read Chapter 18. In Ingmanson & Wallace, read Chapter 19. These are chilling accounts of the human impact on the ocean. The diagrams and photographs are not always easy to look at, but are worthy of careful study.

☐ Watch Program 28: "Marine Pollution."

After Viewing

Review Table 18.2 on page 473 of Garrison, or Table 19.1 on page 421 of Ingmanson & Wallace, for the effect of some common chemical pollutants on health.

Review the sources and effects of lead and DDT. These are both useful chemicals, but DDT production is now severely curtailed. Explain and defend the present ban on their use.

Review the sources of oil in the marine environment. What are some current plans to minimize the effects on areas of rich biologic productivity?

Why are plastics becoming a problem in the marine environment?

Complete the following exercises:

1. The loss of kelp has been a marine disaster. Describe the series of events leading to the decline of the kelp forests. Of what value is kelp to the marine environment and to humans? If you were in charge of a kelp reforestation project, what procedures would you initiate to restore the marine forests?

2. Discuss the problem of assessing the pollution of the oceans at this time. Why is ocean pollution a controversial topic? In fact, why don't we just clean up the sources of pollution?

3. What chemical additives are particularly hazardous and what are their sources? If coal becomes a primary source of energy, will there be any possible environmental problems? See again Table 18.2 on page 473 of Garrison or Table 19.1 on page 421 of Ingmanson & Wallace.

Optional Activities

1. Walk along a shoreline looking for oil or tar on the rocks. What might be some of the sources of the oil? Remember, not all the oil on the rocks is from spills or blowouts from drilling platforms. Try to determine which direction the currents are moving to give you a clue to the location of the source of the oil.

2. Examine Figure 18.2 on page 467 of Garrison, or Figure 19.12 on page 426 of Ingmanson & Wallace. Defend the statement that the supertankers are a clear and present danger to the marine environment and should be banned. Now defend their use, but suggest methods of minimizing the effects of spills, collisions, leaks, and so on that are bound to occur.

3. Walk along the high-tide line of any beach. What plastic objects can you find? Where have they come from? What will be their fate?

Self-Test

1. Additives to the oceans may have which of the following effects on organisms?
   a. They may be lethal to certain organisms.
   b. They may be in the tissues of certain organisms without seriously affecting them.
   c. They may affect only the highest forms in a food chain.
   d. They may benefit certain organisms by providing them with nutrients.
   e. all of the above

2. Scientists now know that Minimata Bay disease is
   a. a vitamin deficiency from a diet of just fish and rice
   b. the cause of fin rot and tumors of fish near outfalls
   c. a form of mercury poisoning from eating contaminated fish
   d. a form of lead poisoning from gasoline fumes
   e. a result of eating contaminated clams during a red tide bloom

3. The loss of the kelp forests has been attributed to
   a. proliferation of kelp-eating urchins
   b. lack of predators to keep the urchin population in ecological balance
   c. sewage sludge coating rocks on which young kelp plants would normally grow
   d. contaminants in the water that are harmful to kelp
   e. all of the above

4. Which of the following statements is true of the communities of benthic organisms near outfall sites?
   a. They have been killed off completely by sludge and heavy metals.
   b. They are subsisting but may consist of animals not usually found in these localities.
   c. They are thriving groups of animals usually found in these localities.
   d. They are providing excellent food fish and sport fishing.
   e. They are living in the kelp forests in great abundance.

5. Petroleum can enter the marine environment from
   a. blowouts and leaks from offshore well sites
   b. natural oil seeps
   c. tanker collisions or wrecks
   d. coastal refineries or leaking storage tanks
   e. all of the above

6. Petroleum in the marine environment has which of the following effects?
   a. It is beneficial to plankton because it shades them from excess sunlight.
   b. It helps marine birds because it oils their feathers.
   c. It adds flavor to food fish that have ingested tar balls.
   d. It is toxic to all wildlife, especially birds and benthic communities.
   e. It provides nutrients to enhance many food chains.

7. Heated waters released by electrical power generating plants
   a. increase the amount of oxygen in the water
   b. improve reproductive success as in the case of the returning salmon
   c. improve metabolic rates and the survival of the species
   d. interfere with breeding cycles and growth rates
   e. all of the above

8. The disappearance of the pelican and other seabirds from the southern California coast was attributed to
   a. the loss of kelp forests, their primary food source
   b. the disappearance of fish in near-shore habitats
   c. oil spills coating the surface waters
   d. DDT in the fish they ate affecting the shells of their eggs
   e. heavy metals in the water affecting the birds' metabolism

9. A new and promising approach to oil pollution will accomplish which of the following?
   a. It will prevent all spills from ever happening.
   b. It will attempt to identify sensitive areas and protect them.
   c. It will prevent tankers from entering all harbors and coastal installations.
   d. It will ban the construction of supertankers.
   e. all of these

10. DDT has begun to increase in the water of some of the coastal regions of the United States. Why is this so?
    a. Increased storm activity in the last few years has stirred up the sediments in which DDT has been suspended, making it available in the near-surface water.
    b. A slight drop in oceanic pH has allowed DDT to redissolve in the water from the sediment.
    c. More is washing off the land, though no new DDT has been added.
    d. It is apparently being added to crops, illegally, by growers wishing to maximize profits at nearly any cost.
    e. It has been formed spontaneously from other pesticides through a previously unknown chemical reaction.

11. An increased amount of carbon dioxide in the atmosphere will
    a. cause an increase in Earth's surface temperature
    b. melt a significant proportion of the Antarctic ice
    c. result in a rise in sea level
    d. cause readjustment of crop-growing zones and rainfall patterns on Earth
    e. all of the above

12. The ultimate environmental insult may well be nuclear war because, according to recent research, even a small nuclear war could have which of the following effects?
    a. make the earth's atmosphere relatively opaque
    b. change weather patterns, triggering storms
    c. result in widespread famine
    d. cause a precipitous drop in surface temperatures
    e. all of the above

Supplemental Reading

Beardsley, T. M., "Not so Hot – New Studies Question Estimates of Global Warming." *Scientific American*, Nov. 1989, 17–18.

Borgese, E. M., "The Law of the Sea." *Scientific American*, vol. 248, no. 3 (1983), 42.

Carson, R., *Silent Spring*. New York: Houghton-Mifflin, 1962. (This is the elegantly written book that is said to have begun the environmental movement in the United States.)

Ehrlich, Anne H., and P. R. Ehrlich, *Earth*. New York: Franklin Watts, 1987. (An excellent overview of environmental issues from two veterans in the field. Beautifully illustrated.)

Ehrlich, P., and C. Sagan, et al., *The Cold and the Dark*. New York: Norton, 1984. (Forecasts a long, cold, somewhat *inconvenient* period following a nuclear war.)

Farrington, J. W., et al., "Ocean Dumping." *Oceanus*, vol. 25, no. 4 (1982–83), 39–50. (A good overview of problems.)

Houghton, R. A., and G. M. Woodwell, "Global Climatic Changes." *Scientific American*, April 1989, 36–44. (Evidence suggests that production of carbon dioxide and methane from human activities has already begun to change the climate and that radical steps must be taken to halt any further change.)

Idyll, C. P, "Mercury and Fish." *Sea Frontiers*, vol. 17, no. 4 (1971), 230–240.

Kerr, R. A., "Stratospheric Ozone Is Decreasing." *Science 239* (1988), 1489–1491. (The article includes a depressing report on the "hole" in the Antarctic ozone layer discovered the previous summer.)

Miller, G. T., *Living in the Environment*, Fifth Edition. Belmont, Calif.: Wadsworth, 1988.

Oceanus, vol. 24, no. 1 (1981). (The whole issue is devoted to pollution and is entitled "The Oceans as Waste Space.")

Oceanus, vol. 20, no. 4 (1977). (This issue, called "Oil in Coastal Waters," discusses the problems of oil pollution.)

Ramanathan, V., "The Greenhouse Theory of Climate Change: A Test by an Inadvertent Global Experiment." *Science 240* (1988), 293–299. (An excellent though technical overview of the greenhouse effect.)

Schell, J., *The Fate of the Earth.* New York: Knopf, 1982. (A closely reasoned and utterly terrifying book on the results of a nuclear exchange.)

Toufexis, A. "The Dirty Seas." *Time*, vol. 132, no. 5 (1988).

Turco, R. P., et al., "The Climatic Effects of Nuclear War." *Scientific American*, vol. 251, no. 2 (1984).

Weisskopf, M., "Plastic Reaps a Grim Harvest in the Oceans of the World." *Smithsonian*, vol. 18, no. 12 (1988).

White, R. M., "The Great Climate Debate." *Scientific American*, July 1990, 36–43. (The author feels that we should take steps to limit the extent of global warming.)

Special issues of Scientific American:

"Managing of Planet Earth," Sept. 1989.

"Energy for Planet Earth," Aug. 1990.

"Meeting the Challenge of Sustainable Development," June 1992.

Lesson Twenty-nine   Hawaii: A Case Study

Overview

Hawaii, the pearl of the Pacific, was missed by the Spanish galleons and not revealed to the Western world until almost three hundred years after Columbus lost his way to the Indies and discovered America. Since the Hawaiian Islands were first seen by Captain Cook in 1778, they have been a magnet, drawing visitors from all over the earth to share in the bounty and beauty of volcanoes in the sea.

The Hawaiian Islands are of particular interest to the oceanographer, for here can be seen a tropical sea with its characteristic climate, winds, currents, and temperatures clearly delineated. The marine organisms of the tropics can also be studied for their complex interactions with one another and with their oceanic island habitat. For these reasons Hawaii is the perfect setting for a case study bringing together the various facets of marine science.

To the tourist, *Hawaii* means one of the major high islands, perhaps only Waikiki Beach in Honolulu on the Island of Oahu. The Hawaiian chain actually consists of over 120 islands and islets extending nearly 3,200 km (2,000 mi) across the middle of the Pacific. Only six attract visitors. The others are uninhabited rocky shoals or coral reefs, the rookeries and nesting sites of countless seabirds. All of the islands are just the peaks of a great volcanic mountain range. Located 3,200 km (2,000 mi) from the nearest continent, they are the most isolated land areas on earth.

Following Captain Cook's discovery, Hawaii was inundated by a host of missionaries, traders, and whalers, all of whom left their mark on the culture and economy of the people. Unfortunately, they also brought disease, and within one hundred years after Cook's arrival 74 percent of the native population had perished.

We have learned, maybe too late, that island ecology is very fragile and poorly equipped to compete with the more vigorous intruders from the outside.

Early scientific exploration of the islands revealed major differences between the larger islands, differences that follow a specific pattern from Hawaii on the southeast to Kauai on the northwest. Hawaii is the least eroded, has the least amount of sod and the fewest plants, and stands the highest above the sea. Mauna Loa at 4,170 m (13,680 ft) and Mauna Kea at 4,202 m (13,784 ft) are almost 9,146 m (30,000 ft) above their bases on the deep ocean floor, which makes them actually taller than Mt. Everest!

The only active volcanoes in the chain are Kilauea and Mauna Loa, both on the Big Island of Hawaii. Kilauea and its associated vents and craters have been erupting sporadically for decades and are considered to be in an active phase. Flank eruptions of Kilauea have been extending the island on the south. Just offshore, a small sub-sea volcano is actively growing, also adding to the size of Hawaii. Except for a small disturbance in 1975, Mauna Loa had been dormant for over 30 years. The last major eruption occurred in 1950. A few years ago, seismographs on the slopes and summit began to pick up rumblings that indicated molten rock was moving again within the magma chamber. On March 25, 1984 Mauna Loa erupted, lighting up the sky and flooding the summit and some of the slopes with basalt lavas characteristic of oceanic island volcanoes.

Hawaiian eruptions are considered "quiet" as compared with the explosive volcanoes of the island arcs. During the Mauna Loa eruptions, there were no injuries, very little damage to structures, and no tsunami were set in motion. Will this active period continue, or will Mauna Loa once again become dormant? Was this the last gasp of a dying vent? Only time will tell.

Hawaii is, therefore, the youngest of all islands, and probably began to form less than one million years ago.

Going northwestward, the next island is Maui. The summits are lower, about 3,049 m (10,000 ft) above sea level. Maui is older and has not erupted since 1790. The island is deeply eroded, and the basalt rocks have weathered into the red soils that support acres of sugar cane and pineapple.

The pattern is now beginning to develop. On Oahu, the next major island to the northwest, the original slopes of the volcano have been completely eroded away. The rocks are older still, and the sods are deeply weathered. These volcanoes have not erupted in historic times. Diamond Head, the great landmark of the Pacific, was standing there to greet the first Polynesians who arrived from Tahiti about 500 A.D.

Continuing on to Kauai, the great canyon of Waimea gives evidence of the extensive and lengthy erosion that has shaped this island. The name Garden Isle aptly describes the lush vegetation that thrives in the deep red soil. The Hawaiian Island chain does not end here, however, for the line of small reefs and shoals continues northwesterly to join the Emperor chain of seamounts near the great bend beyond Midway Island. Volcanic activity ceased on Midway about 20 million years ago. The Emperor chain strikes due north and heads into the junction of the Kurile Trench and the Aleutian Trench in the North Pacific. This chain is a series of seamounts rarely breaking the surface of the sea. The most ancient volcanoes in the archipelago, these seamounts formed about 70 million years ago. Since that time they have been planed off by waves and remain as sunken survivors of what might have been the mightiest peaks of them all.

Early oceanographers recognized that this pattern of volcanism must be significant to the origin of both the Emperor seamounts and the Hawaiian Islands. But it wasn't until the advent of plate tectonics that an acceptable explanation accounted for the varying character of the islands. We now know that the Pacific plate is moving northwestward, moving away from the East Pacific Rise, sliding along the San Andreas Fault on the east, and subducting into the trenches in the north and west. We also believe that a hot spot exists in the mantle or lower crust where molten rock rises through the ocean crust

to flow out on the seafloor. The continued eruption of lava piles flow upon flow, building a cone that in the course of a million or so years breaks through the sea surface. The volcano continues to grow over the hot spot, bit by bit constructing an island in the shield shape characteristic of basalt volcanoes. Kilauea, possibly Mauna Loa, and the south end of Hawaii are over the hot spot at the present time.

The crustal plates continue to move, however. As the seafloor and the island are rafted slowly away from the hot spot they become decoupled from the rising magma. The active vents become dormant and finally extinct. Erosion now becomes the dominant force. Streams cut into the basalt carving deeply into the very heart of the cold crater. The summits are slowly lowered; the shield shape may be completely destroyed as on Oahu; the old vents become hardly discernible. The basalt rocks weather slowly forming the red soils so advantageous to plant growth. The waves, too, eat away the land until the island is only a small fraction of its original size. Coral reefs start to fringe the shores, a new ecosystem forms on the rocky substrate, and a platform of living organisms builds out around the old island.

Meanwhile, back at the hot spot, the magmas rise again, punch through the seafloor that has now moved over the hot spot, and again the cycle starts leading to formation of another volcano in the sea. And so each seamount and island of the Emperor and Hawaiian chains is born over the hot spot, grows to its appointed size, and moves on. Where will the new island form if our theories are correct?

Today, the Hawaiian Islands are in the belt of the tropics and the climatic conditions are dominated by the ever-blowing trade winds. The seasonal variations on the islands are less than the variations from the windward to lee sides, and less than the changes encountered going from the shores to the island summits. The east and northeast sides of the islands face into the trades and are subject to attack by the large waves driven in by the winds. The windward coasts are frequently steep, cliffed, or rocky, and the high-energy beaches lack sand. The rains, too, come in on the windward side as the trades pick up abundant moisture from the ocean between the continent and the islands. When these moist winds strike the land they are lifted onto the cool upper slopes. The moisture in the air condenses and may drench the summits with as much as 400 in. of rain per year!

The lee sides away from the trades receive very little rain and may be hot, dry, and almost windless. Vegetation characteristic of desert regions and cactus are not uncommon sights in some areas. The lava slopes show little weathering or erosion, and there are few canyons or deep valleys on the lee side. Even the old lava flows may have a surprisingly fresh appearance.

The oceanic environment surrounding the Hawaiian Islands is also affected by its location within the belt of the tropics, as well as by the persistent trade winds. The waters around Hawaii are warm and stay near the 77°–80°F mark all year. The temperatures are a little cooler here than the open ocean waters to the west because of the influence of the cold California Current that feeds the North Equatorial Current flowing near the islands. The salinity is a little lower than the open waters to the north and west, again influenced by the low-salinity California Current, but also by excessive precipitation around the islands.

The major surface currents driven by the trades come from the east. Small eddies and currents wheel around the islands and race through the deep straits between them. In general, the islands are too small to develop strong longshore currents and upwellings are rare.

The tides are not particularly notable in Hawaii, with the maximum ranges less than 0.06 m (2 ft). The intertidal zone is small, and the giant kelps and other large seaweeds of the cooler waters are lacking completely.

The beaches, of course, are incomparable. Some beaches are covered with black sand, bits of black basalt, or black volcanic glass called obsidian, which formed when the hot lava flowed into the cold water. Others are green, the sands composed of clear, glassy green grains of olivine, an abundant volcanic mineral. The red sand beaches of red lava cinders are not common but, combined with the blue sea and green plants, they make a beautiful and startling sight.

The white sands of many famous Hawaiian beaches are organic rather than volcanic in origin. The particles are derived from skeletons of coral or calcareous algae, from the shells or hard parts of a variety of reef dwellers, torn from the reef by the waves and washed up on the shore. These sands viewed under a microscope are a mosaic of tiny perfect shells in a variety of shapes and colors.

The carbonate reef communities just beyond the white sand beaches lure divers, from novice snorkler to expert scuba diver, not so much for the food fish or occasional lobster but for a glimpse of one of the most beautiful and complex ecosystems on earth today. The colorful fish; the red, black, purple, and green spiny sea urchins; and the intricate architecture of the coral heads are but a small part of this tropical community. The clarity of the water adds to the enjoyment but, of course, does not add much to the nutrient requirements of the reef inhabitants. Highly specialized relationships; unusual ways of making a living; and almost desperate measures to find food, locate a mate, and avoid being eaten are all part of life in this colorful, but barren environment.

In spite of the great variety of species living here, we now understand that brilliant blue water signals a biologic desert. Only the efficiency of the reef community and tight recycling of organic material allow the ecosystem to survive and increase in size.

Some reefs are barely surviving, however, because of the pressures of urban development and the dumping of raw sewage into bays. Kaneohe Bay on the windward side of Oahu has been seriously depleted of life, and efforts are underway to improve this dismal situation.

Beyond the reef the seas do not exactly teem with life, but the large tunas are there preying on smaller fish. The green, blue, and gold dolphin fish called mahi-mahi are rightfully prized for their excellent flavor and therefore are strenuously hunted. Marlin and other bill-fish lure fishermen to the Kona coast for some of the world's finest sport fishing. Sharks are not strangers to tropical waters but are rarely seen around the reefs or beaches. Sea snakes are also rare, which is fortunate because they are poisonous. Sea turtles come close to shore occasionally, bobbing in the water as they warily eye the snorkelers. The most spectacular marine animals in the waters near Hawaii are the humpback whales that sing, play, and probably breed off west Maui. Whaling is no longer permitted as it once was in Hawaii, but whale watching is, as tourists flock in to get a close look at the huge marine mammals.

From our study of this group of distant islands we will see how the moving crustal plates, the hot spot, the trade winds, and the tropics all contribute to the physical and biological environment that characterizes the Hawaiian Islands.

Learning Objectives

After completing the reading assignment and viewing the program, the student will be able to:

☐ Discuss the hot spot theory of the origin of the Hawaiian Islands, and present supporting evidence.

☐ Compare and contrast the islands of the Hawaii chain in terms of age, active volcanism, height of the peaks, erosion, soil development, and plants. (Start at the Big Island of Hawaii and travel northwestward to Kauai.)

☐ Discuss the effects of the trade winds in Hawaii, comparing and contrasting the windward and lee sides as to climate, erosion, and topographic features.

☐ Discuss the sand beaches of Hawaii, contrasting the major kinds of sediments, their appearance, and origin.

☐ Describe the currents, temperature, and life of the tropical seas near Hawaii.

Key Terms and Phrases

Hot spot A persistent heat source in the asthenosphere that generates the molten magma that supplies certain seafloor volcanoes. There is good evidence that chains of volcanoes are the result of motion of the oceanic plate over a hot spot. Today, the active volcano Kilauea on the Big Island of Hawaii is believed to be located over a hot spot in the middle of the Pacific plate.

Shield shape The profile of volcanoes built of fluid basalt lavas, in contrast with the steep slopes of volcanoes constructed of cinders and solid particles. The gentle slopes resemble the shields of ancient warriors. The shield shape is characteristic of the Hawaiian volcanoes.

Kilauea Iki A small pit crater near the summit of Kilauea, the site of a series of spectacular eruptions in 1957–59. The lavas and cinders built a cinder cone on the summit, denuded a

forest of ohia trees creating the "devastated area," and later destroyed the small farming village of Kapoho on the south flank of the volcano.

Before Viewing

☐ Specifically review Program and Lesson 25: "The Tropic Seas."

☐ Reread pages 82–84 of Garrison, reviewing the "hot spot" theory, and noting especially Figure 3.30 showing the complete Hawaiian chain from space. Review the material on the colonization of Hawaii on pages 29–31. If you are reading the text by Ingmanson & Wallace, review page 58 in Chapter 4, and pages 73–77 in Chapter 5, which discusses the "hot spot" theory of the origin of the Hawaiian Islands.

☐ Review the illustrations in either text describing the coral reef habitat and inhabitants.

☐ Watch Program 29: "Hawaii: A Case Study."

After Viewing

Review the text sections and the Overview being sure you understand the key terms and phrases, and complete the following exercises:

1. You are driving from the windward side of Hawaii to the lee side. Describe the changes in the weather and climate, vegetation, and topography, and discuss the factors contributing to these differences. On which side, windward or lee, will you find better swimming, and why?

2. You have decided to drive from the beach to the summit area of Kauai. Describe the changes you will see on this excursion. What causes these variations?

3. You are on an extended tour of the major high islands, starting at the southeastern end on the Big Island of Hawaii and ending at the northwestern end, Kauai. Describe some of the differences you have noted as you proceeded from island to island. What patterns do you discern? How do we explain

these patterns according to modern plate tectonic theory?

4. As you travel around the islands, notice the different colors of the beaches. Describe the colors you see and the kind of deposits you find on each beach.

Optional Activities

1. Using Figure 3.30 (page 83) or the map on page 72 of Garrison, or after the Index at the end of Ingmanson & Wallace, locate the Hawaiian Islands. Trace the chain northwestward to the Emperor seamounts. The change in direction of the two chains of volcanoes is a clue to something that must have happened to the Pacific plate, but what? (The plate changed direction of movement. Why, we don't know. In fact, we still don't know exactly why the plates move at all!)

2. If you have been to Hawaii, take out your slides and pictures and try to identify the windward and lee sides of the islands, the beaches, and the summits. In other words, look at the scenes again but with the analytical eye of the oceanographer.

3. If you haven't been to Hawaii yet, look at pictures in travel books or in newspapers and try to analyze the dominant factors operating.

4. Take another swim on the coral reef on pages 426–427 of Garrison, or pages 354–355 of Ingmanson & Wallace. Watch the triggerfish (no. 6) as it feeds. Notice where the eyes are – far back on the head. The triggerfish is perfectly adapted for preying on prickly crustaceans and sea urchins. The black dots on the tail are rows of sharp spines that are used in sideswiping or tail-lashing an enemy Do you see the moray eel? Although it looks ferocious, eels are shy and try to stay hidden in crevices and holes in the reef. The eel is an efficient night hunter with an acute sense of smell. The eel in the picture is being "cleaned." Can you see the cleaner?

Self-Test

1. Volcanic activity in the Hawaiian Islands can be found
   a. on every major high island
   b. on the northwestern island of Kauai
   c. in the Emperor Chain north of Hawaii
   d. in central and southern volcanoes on the Big Island of Hawaii
   e. None of the above. Volcanic activity has not occurred anywhere since 1790.

2. The volcanic vent presumably located over the hot spot today is
   a. the summit crater on Kauai
   b. Diamond Head on Oahu
   c. Kaneohe Bay on Oahu
   d. Haleakala, the summit of Maui
   e. Kilauea and Mauna Loa on the Big Island

3. As we study the Hawaiian chain, we notice as we travel from south to north which of the following changes occur(s)?
   a. The islands get progressively older.
   b. The soils get more weathered and deeper.
   c. Erosion has carved deeper canyons and valleys.
   d. The summits are lower.
   e. All of the above are correct.

4. The windward sides of the islands are characterized as
   a. hot and dry
   b. covered with desert vegetation including cactus
   c. covered with lush green vegetation and deeply eroded canyons
   d. areas with little soil and relatively unweathered lava rock
   e. quiet beaches with small waves

5. The Hawaiian Islands are located in the windbelt of the
   a. prevailing westerlies
   b. northeast trade winds
   c. southeast trade winds
   d. polar easterlies
   e. variable winds that change with the seasons

6. The water temperatures near Hawaii are a little cooler than the oceans to the west due to
   a. a windchill factor
   b. rather severe winters on the islands
   c. strong upwellings of cold water
   d. the influence of the cold California Current on the North Equatorial Current
   e. the low angle of the sun at that latitude

7. The tropical open ocean surrounding the Hawaiian Islands
   a. is teeming with plankton
   b. supports great kelp forests
   c. is murky or turbid with abundant dissolved nutrients
   d. is clear blue and generally lacking nutrients and plankton
   e. is the most biologically productive water on earth

8. The famous black sand beaches of Hawaii are composed of
   a. particles of basalt and obsidian
   b. shells of mollusks and sea urchin spines
   c. bits of coral skeleton and calcareous algae
   d. grains of olivine
   e. quartz and feldspar grains

9. Which of the following statements is characteristic of the range of tides in Hawaii?
   a. It is generally greater than on mainland shores.
   b. It has produced a well-marked intertidal zone.
   c. It has resulted in zonation of both kelps and intertidal animals.
   d. It is very low, usually less than 2 ft.
   e. It is variable and may cause damage to fringing reefs by exposing the organisms to the hot sun.

10. An enthusiastic group of tourists now flock to Maui in the winter
    a. to enjoy the newly opened ski resorts
    b. to see the eruption of Haleakala on Maui
    c. to see the humpback whales in the waters of west Maui
    d. to watch the training of mahi-mahi in Sealife Park
    e. for the free pineapple juice

Supplemental Reading

Peck, D., T. L. Wright, and R. W. Decker, "The Lava Lakes of Kilauea." *Scientific American*, vol. 241, no. 4 (1979), 114–28.

Dvorak, John J., Carl Johnson, and Robert I. Tilling. "Dynamics of Kilhauea Volcano." *Scientific American*, vol. 267, no. 42 (1992).

Lesson Thirty Epilogue

Overview

The marine sciences are today at the threshold of a new age. Recent revolutions in biology and geology are being assimilated, and the road ahead seems clearer. The whole approach to oceanography is changing rapidly. Satellite-borne sensors can provide data in an instant that would have taken years to collect using surface ships. Even the use of ships has changed. Scientists now have the ability to make many direct measurements simultaneously, frequently depending on interagency or even international consortia. Measurements such as these, taken from different ships in various locations at the same time, are called synoptic surveys.

By means of magnetometric and reflected sound technology the continental shelves of the world are being surveyed for energy and mineral resources. Most of the recoverable petroleum on which our energy-intensive society will depend until alternative sources can be developed, lies under the surface of the seas. New recovery methods are also being developed to safely deliver this oil and natural gas to processing centers. Mineral beds have been found in the deep ocean basins. The mechanisms needed to mine these areas are now being perfected. Even the complex internal structure of the ocean is yielding to computerized analyses similar to those developed for tomography in human medical studies.

Ocean food gathering will eventually be replaced by large-scale ocean farming. This undertaking is not without difficulties. The wet pasture of the ocean is constantly in motion; the animals may not be easily contained by fences. Some scientists estimate that up to 90 percent of the nutrients in the sea are trapped in deep waters where they cannot be used by plants because of the lack of sunlight. Perhaps artificial upwelling, especially in the tropical areas of the world, could greatly increase food supplies provided by the ocean. Mariculture today remains an expensive and energy-consuming way to produce luxury seafoods, such as abalone, lobster, and shrimp.

Water itself is an important oceanic resource. Water could be transported in frozen form from the Arctic or Antarctic regions to provide relief to desert coastal areas including the Middle East and southern California. Water could be (and is) distilled from the ocean using low-pressure evaporators in many parts of the world.

Experimental electrical power schemes harness the temperature differential between surface water and deep ocean water. By selecting the correct working fluids and internal equipment, a power plant could theoretically be operated from the ocean's thermal differential.

Medicine and pharmacology, have advanced due to discoveries from the ocean. Food sciences also hold the ocean in their debt. There are very few areas of human endeavor that have not benefited from oceanographic science. These advances have not been purchased at low cost. Often the price is more than financial – the environmental consequences are disturbing and may harbor negative implications of which we have not yet dreamed. After all is said, though, it may be that the ocean's greatest permanent gift to humanity is intellectual – the constant ever-changing challenge its inspiring mass presents. The writers of this guide find this challenge to be a very personal and compelling thing. *We want to know* about the ocean. We haunt our mailboxes for journals, search TV listings for any new ocean shows, hungrily inspect new rocks or organisms with the enthusiasm of little kids, delight in the insignificant amount of knowledge we already have, and share our insight with anyone at the drop of a hat. We are personally delighted that you have come with us this far.

All of us stand as the Greeks, the Mediterranean immediately in our wake, sailing into a vast and unknown world. What we do *with* and *to* the ocean is literally of planetary consequence. In the last century we have developed the physical, chemical, and biological machinery to destroy or rejuvenate the world ocean and all of its life. Intelligence and beauty must triumph; we have no other rational alternative. We hold an ocean within us.

Learning Objectives

After completing the reading assignment and viewing the program, the student will be able to:

☐ List some of the directions for oceanographic research in the next decades.

☐ Recognize the diversity of motivations expressed by ocean scientists.

☐ Discuss some of the broad consequences of oceanic mismanagement.

Key Terms and Phrases

Synoptic survey A survey in which simultaneous measurements are taken from a variety of platforms in a number of different locations.

Mariculture Sea farming. Aquaculture includes both sea farming and freshwater farming.

Magnetometer A device that measures subtle changes in the strength and direction of the earth's magnetic field.

Before Viewing

☐ Take this opportunity to review the material in your textbook. Look to see where you have been. If you are using the Garrison text, please read the Afterword on pages 490–491.

☐ Watch Program 30: "Epilogue."

After Viewing

Complete the following exercise:

☐ List three projects in ocean research that you would like to see initiated at the federal level. Why? What would you hope would be the results of your proposals? Would they be worth the cost?

Optional Activities

Now that you have completed the reading and watching for this course, have tried the exercises, have become involved in some of the optional activities and supplemental reading, have engaged in some quiet learning about the oceans, and have applied some of this information to the "real world," ask yourself these questions: How have you changed? Did you find anything especially surprising or hard to believe? Did anything you learned contradict your previous understanding of some ideas? What did you find most memorable and exciting about this series? Has this series fulfilled your expectations?

The oceans may be appreciated at many different mental levels. Those of us who enjoy an oceanographic background (and this now includes you) look at the ocean with a greater understanding than we did before we began. We find pleasure in knowing things about it. We love to talk to others about its many mysteries and pleasures. The whole concept of an ocean world appeals to us. *Welcome to our minds and thoughts.* We hope you continue with our science and appreciate its benefits and its limitations.

Answer Key for the Self-Tests

Lesson One: The Water Planet

1. e	4. a
2. b	5. c
3. a	6. a

Lesson Two: Cosmic Origins

1. c	6. b
2. c	7. d
3. a	8. e
4. c, d	9. b
5. b	

Lesson Three: Historical Perspectives

1. e	8. d
2. a	9. e
3. c	10. b
4. c	11. e
5. e	12. c
6. d	13. b
7. e	

Lesson Four: The Waters of the Earth

1. b	7. d
2. e	8. e
3. c	9. d
4. d	10. b
5. d	11. b
6. a	12. c

Lesson Five: Ocean's Edge

1. c	7. d
2. b	8. c
3. e	9. e
4. c	10. d
5. b	11. d
6. e	12. b

Lesson Six: The Intertidal Zone

1. a	6. d
2. b	7. b
3. e	8. a
4. e	9. c
5. c	10. c

Lesson Seven: Continental Margins

1. a	7. c
2. e	8. b
3. d	9. d
4. d	10. e
5. a	11. b
6. e	12. e

Lesson Eight: Beyond Land's End

1. e	7. a
2. d	8. d
3. c	9. b
4. d	10. e
5. c	11. d
6. c	

Lesson Nine: Plate Tectonics

1. e	7. d
2. b	8. c
3. e	9. e
4. d	10. c
5. c	11. d
6. b	

Lesson Ten: Islands

1. d	6. d
2. a	7. c
3. e	8. a
4. c	9. b
5. e	10. b

Lesson Eleven: Marine Meteorology

1. e	8. d
2. d	9. a
3. d	10. e
4. e	11. b
5. b	12. e
6. b	13. e
7. c	14. b

Lesson Twelve: Ocean Currents

1. b	6. b
2. e	7. d
3. c	8. a
4. a	9. b
5. b	10. e

Lesson Thirteen: Wind Waves and Water Dynamics

1. b	7. e
2. d	8. d
3. e	9. c
4. d	10. e
5. a	11. c
6. b	12. b

Lesson Fourteen: The Ebb and Flow

1. c	6. d
2. d	7. c
3. b	8. e
4. a	9. a
5. e	10. d

Lesson Fifteen: Plankton: Floaters and Drifters

1. a, b	7. b
2. a, c, d	8. c
3. a, c, e	9. e
4. a, b	10. d
5. d	11. c
6. a	

Lesson Sixteen: Nekton: Swimmers

1. d	7. b
2. e	8. c
3. a	9. b
4. e	10. b
5. e	11. c
6. a	

Lesson Seventeen: Reptiles and Birds

1. e	5. e
2. d	6. d
3. c, d, e	7. b
4. d	8. d

Lesson Eighteen: Mammals: Seals and Otters

1. c	5. c
2. a	6. a
3. c, e	7. d
4. e	8. a

Lesson Nineteen: Mammals: Whales

1. d	8. e
2. b	9. e
3. b	10. e
4. e	11. b
5. c	12. c
6. e	13. b
7. a	14. e

Lesson Twenty: Living Together

1. d	6. b
2. d	7. c
3. e	8. e
4. a	9. a
5. b	10. c

Lesson Twenty-one: Light in the Sea

1. c	7. d
2. e	8. e
3. e	9. a
4. e	10. c
5. d	11. d
6. b	

Lesson Twenty-two: Sound in the Sea

1. a	7. e
2. b	8. e
3. e	9. d
4. d	10. e
5. c	11. b
6. d	

Lesson Twenty-three: Life Under Pressure

1. d		6.	d
2. c		7.	e
3. c		8.	c
4. a		9.	d
5. e		10.	d

Lesson Twenty-four: The Polar Seas

1. e		6.	b
2. a		7.	c
3. b		8.	d
4. a		9.	c
5. d		10.	a

Lesson Twenty-five: The Tropic Seas

1. c		7.	d
2. e		8.	c
3. e		9.	b
4. b		10.	a
5. d		11.	e
6. e			

Lesson Twenty-six: Mineral Resources

1. e		6.	e
2. d		7.	b
3. c		8.	d
4. e		9.	e
5. e			

Lesson Twenty-seven: Biological Resources

1. b		6.	e
2. d		7.	d
3. e		8.	c
4. c		9.	d
5. e		10.	e

Lesson Twenty-eight: Marine Pollution

1. e		7.	d
2. c		8.	d
3. e		9.	b
4. b		10.	d
5. e		11.	e
6. d		12.	e

Lesson Twenty-nine: Hawaii: A Case Study

1. d		6.	d
2. e		7.	d
3. e		8.	a
4. c		9.	d
5. b		10.	c